烹饪专业及餐饮运营服务系列教材

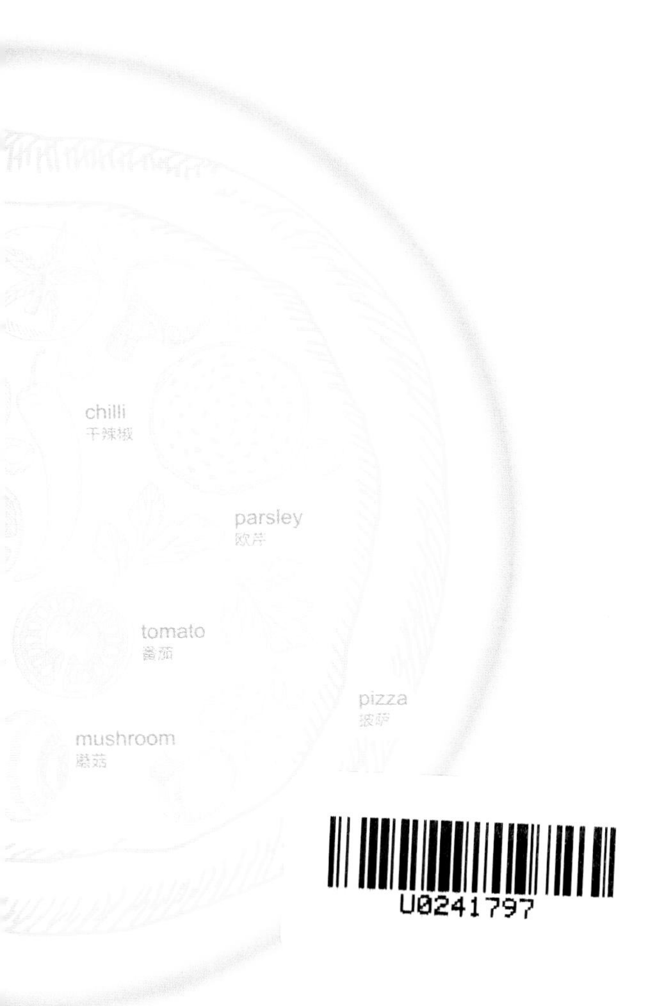

WESTERN COOKING ENGLISH

西餐烹饪英语

（第5版）

主　编　蔡琳琳

副主编　熊梓彤　黄晨曦　倪　楠　殷佳妮

旅游教育出版社

·北京·

图书在版编目（CIP）数据

西餐烹饪英语 / 蔡琳琳主编. -- 5版. -- 北京 : 旅游教育出版社, 2023.1
烹饪专业及餐饮运营服务系列教材
ISBN 978-7-5637-4512-8

Ⅰ. ①西… Ⅱ. ①蔡… Ⅲ. ①西式菜肴－烹饪－英语－中等专业学校－教材 Ⅳ. ①TS972.118

中国版本图书馆 CIP 数据核字(2022)第 237476 号

烹饪专业及餐饮运营服务系列教材
西餐烹饪英语
（第5版）
主编 蔡琳琳
副主编 熊梓彤 黄晨曦 倪 楠 殷佳妮

策　　划	景晓莉
责任编辑	景晓莉
出版单位	旅游教育出版社
地　　址	北京市朝阳区定福庄南里1号
邮　　编	100024
发行电话	(010)65778403　65728372　65767462(传真)
本社网址	www.tepcb.com
E-mail	tepfx@163.com
印刷单位	唐山玺诚印务有限公司
经销单位	新华书店
开　　本	787毫米×1092毫米　1/16
印　　张	11.5
字　　数	144千字
版　　次	2023年1月第5版
印　　次	2023年1月第1次印刷
定　　价	49.80元

前言

Preface

此教材再版之际，正值中国共产党第二十次全国代表大会胜利闭幕之时。

为贯彻落实党的二十大精神，加快推进党的二十大精神进教材，进课堂，进头脑，我社对《西餐烹饪英语》教材进行了及时修订。

二十大报告指出，要推进高水平对外开放，推动共建一带一路高质量发展。涉外旅游行业是中国对外开放的窗口，是面向世界传播中国文化、讲好中国故事、服务好世界友人的重要渠道。高素质涉外旅游服务人才是推动中国涉外旅游及贸易发展的重要支撑。

为充分发挥旅游业服务国家“高水平对外开放”的功能和作用，响应国家从以制造业为主的开放扩展到以服务业为重点的开放政策，我社将教材的编写与开发重点放在了培养面向高水平对外开放的旅游服务人才上，先后开发了《西餐制作》《西式面点制作》《西餐原料与营养》《热菜制作》《冷菜制作与艺术拼盘》《食品雕刻》《酒水服务》《西餐服务》《西餐烹饪英语》《饭店服务情境英语》《导游讲解》《旅游服务礼貌礼节》《旅游概论》等中等职业教育外向型专业课精品教材。

中西餐制作及与之相关的西餐服务、酒水服务、旅游服务等外向型教材的配套开发，有助于实现整个产业链与复合型人才培养模式的紧密对接，有助于引导读者从服务的角度审视菜品制作，从高水平对外开放的高度理解自己将要从事的职业，初步建立起菜品制作与旅游服务相互补充的知识体系，能够用发展的眼光、互联互通的思维看待旅游服务业。

《西餐烹饪英语》为中等职业教育西餐烹饪专业配套英语教材，首版教材于 2013 年出版，本版为第 5 版。教材秉承做学一体能力养成的课改精神，适应项目学习、模块化学习等不同学习要求，注重以真实生产项目、典型工作任务等为载体组织教学单元。其既可作为中职院校学生的专业核心课教材，也可作为岗位培训教材。

教材分为5个模块共23个单元，内容包括学生自我介绍、西厨设备工具、西餐主要烹饪方法、主要菜品、主要调味品及香料等西餐厨房基础知识专业英语，汁、汤、蔬菜、土豆类食物、米饭类食物、牛肉、猪肉、羊肉、鱼类和贝类、禽类等开胃菜和主菜烹制专业英语，意大利面食、面包、蛋糕、巧克力和姜饼等西式面点制作常用专业英语，以及星级酒店西餐菜单、西餐服务情景对话等多个专题。每一单元精心设计了四个教学环节，让学生依次通过看图、连线、认菜品说原料及读单词，对西餐烹饪原料和菜品等专业英语术语进行循序渐进的学习。

与同领域其他教材相比，本教材可谓另辟蹊径，其融合了重复记忆与有效的联想词汇学习策略，通过精选的西餐菜品和典型菜例，强化专业英语学习。学生在专业课上亲手制作菜品，回到英语课堂中及时学习对应菜品的原料词汇及相关英语知识，通过不断重复记忆，达到记住并运用专业词汇的目的。

为提升读者的阅读体验感，此版由单色改为彩色印刷。

本教材由桂林市旅游职业中等专业学校创作团队编写；蔡琳琳任主编，黄晨曦撰写Module1，倪楠、黄晨曦撰写Module2，蔡琳琳撰写Module3，熊梓彤撰写Module4，熊梓彤、黄晨曦撰写Module5；部分菜点图片由西餐烹饪专业教师殷佳妮提供，旅游教育出版社景晓莉在原有图片基础上进行了大量增补和修图工作；二维码教学资源由旅游教育出版社制作；上海茂盛山房行政总厨潘熠林先生校正了相关专业词汇，王蓉、李秀慧、关星、张蓓、王晓楠校正了词汇及音标。

在编写本书的过程中，编写团队得到了桂林市旅游职业中等专业学校西餐烹饪专业老师顾健平的悉心指导，同时得到了桂林香格里拉大酒店行政总厨罗明先生、送餐部经理覃琦棋女士的真诚帮助，在此表示衷心感谢。

由于编者水平有限，书中难免存在错漏之处，恳请各位专家、读者批评指正。

旅游教育出版社

2023年1月

目录
Contents

Module 3

Module 4

Module

模块 1

Unit 1

Self-introduction 自我介绍

词汇在线

My name is ×××.	我的名字叫 ×××。
I was born on...	我出生于……(具体出生日期)
I was born in...	我出生在……(出生月/年份、地点)
I am 16 years old.	我今年 16 岁。
I major in western-style food cooking specialty.	我主修西餐烹饪专业。
I study in Guilin Central Vocational School.	我在桂林职业教育中心学校学习。
I graduated from Guilin Tourism Vocational Secondary School.	我毕业于桂林旅游职业中等专业学校。
My favorite sport is playing table tennis.	我最喜欢的运动是打乒乓球。
I like surfing the internet, listening to the music, playing football, singing, and dancing.	我喜欢上网、听音乐、踢足球、唱歌和跳舞。
I got the first/second/third prize in English Songs Competition.	我在英语歌曲比赛中获得了一等奖、二等奖、三等奖。

Words List

主修	major ['meɪdʒə] v.
西方的；欧美的	western ['west(ə)n] adj.
方式；样式	style [staɪl] n.
专业，专长	specialty ['speʃ(ə)ltɪ] n.
旅游；旅游业	tourism ['tʊərɪz(ə)m] n.
中央的；中心的；主要的	central ['sentr(ə)l] adj.
职业的；行业的	vocational [və(ʊ)'keɪʃ(ə)n(ə)l] adj.
毕业；接受学位	graduate ['grædʒʊət; -djʊət] v.
中等的；第二的；次要的	secondary ['sek(ə)nd(ə)rɪ] adj.
最喜爱的	favorite ['feɪvərɪt] adj.
乒乓球；桌球	table tennis ['teɪb(ə)l 'tenɪs]
冲浪；在（因特网）上浏览	surf [sɜːf] v.
因特网；互联网	internet ['ɪntənet] n.
奖赏；奖金；奖项	prize [praɪz] n.
比赛；竞赛	competition [ˌkɒmpɪ'tɪʃ(ə)n] n.

词汇在线

Unit 2

Various Style Dishes 各式各样的菜品

- style [staɪl] n. 方式；样式
- dishes ['dɪʃiz] n. 盘（dish 的名词复数）；一盘食物

❶ French Style 法式

French-style dishes　法国菜
French [fren(t)ʃ] adj. 法国的

法国国宝级菜式——焗蜗牛

❷ British Style 英式

British-style dishes　英国菜
British ['brɪtɪʃ] adj. 不列颠的；英国的

三明治的故乡——英国

❸ American Style 美式

American-style dishes　美国菜
American [ə'merikən] adj. 美国的

西式快餐的主角——美式汉堡

❹ Russian Style 俄式

Russian-style dishes　俄国菜
Russian ['rʌʃ(ə)n] adj. 俄国的，俄罗斯的

俄式菜中的贵族——鲟鱼子酱

❺ Italian Style 意式

Italian-style dishes 意大利菜
Italian [ɪ'tæljən] adj. 意大利(人)的

意式肉酱面——经典中的经典

❻ German Style 德式

German-style dishes 德国菜
German ['dʒɜːmən] adj. 德国(人/语)的

德国国宝——香肠

❼ Austrian Style 奥式

Austrian-style dishes 奥地利菜
Austrian ['ɔstriən] adj. 奥地利(人)

遍布香肠摊的奥地利

❽ Hungarian Style 匈式

Hungarian-style dishes 匈牙利菜
Hungarian [hʌŋ'g əriən] adj.匈牙(人)的

红椒粉——匈牙利菜的灵魂

Tips **翻译英文菜品名称的方法**

烹饪方法（动词过去分词+主料+with+配料）

词汇在线

Unit 3

Equipment and Tools 设备与工具

I Kitchen Range 灶

- kitchen ['kɪtʃɪn; -tʃ(ə)n] n. 厨房
- range [reɪn(d)ʒ] n. 范围；类别

❶ gas stove 火灶

gas [gæs] n. 气体；[矿业] 瓦斯；汽油
stove [stəʊv] n. 炉，火炉

❷ baking oven 烤箱

baking ['beɪkɪŋ] adj. 烘烤的
oven ['ʌv(ə)n] n. 烤箱，烤炉

❸ microwave oven 微波炉

microwave ['maɪkrəʊweɪv] n.
微波；微波炉

❹ salamander oven 明火炉

salamander ['sælə,mændə] n.
烤箱；火蜥蜴，火怪，耐火的人

II Machines 机器类

◆ machine [mə'ʃi:n] n. 机器，机械

❶ refrigerator 冰箱

refrigerator [rɪ'frɪdʒəreɪtə] n.
冰箱；冷藏库

❷ waring blender 立式万能机；捣碎机

waring ['weəɪŋ] n.华林；华陵（姓氏）
blender ['blendə] n. 掺和器，搅拌机

❸ egg beater 打蛋机

beater ['bi:tə] n. 拍打器；敲打者；搅拌器；打浆机

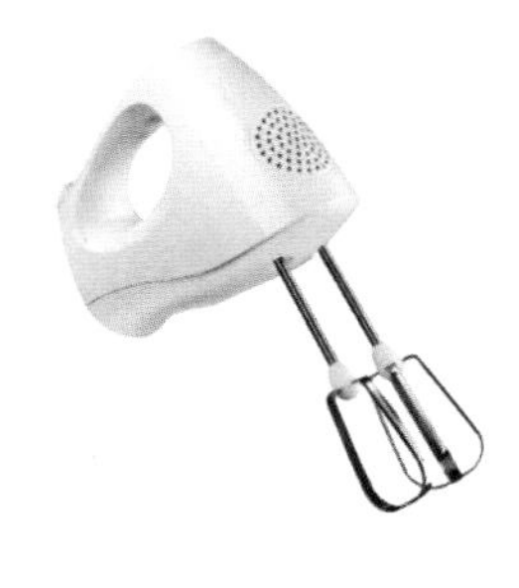

❹smashing appliances 多功能粉碎机

smashing ['smæʃɪŋ] adj.粉碎的
appliances [əp'laɪənsɪz] n. 器具，装置

❺ slicer 切片机

slicer ['slaɪsɚ] n. 切薄片的机器；切刀

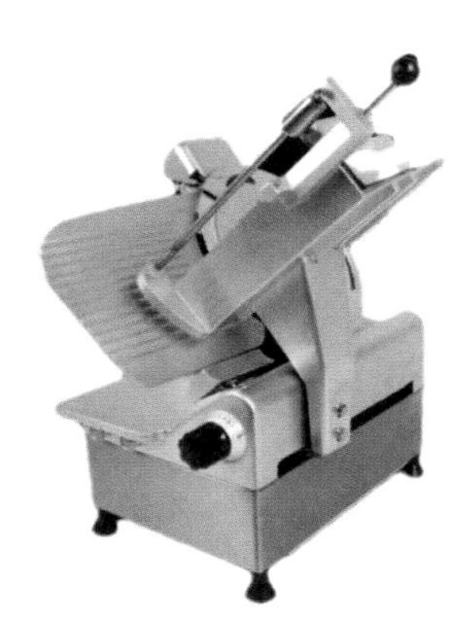

III Cooking Pots 锅

- cooking ['kʊkɪŋ] adj. 烹调用的(水果、锅、炉等)
- pots [pɔts] n. 罐(pot 的名词复数)；(某种用途的)容器

❶ stockpot 汤锅

stockpot ['stɒkpɒt] n.　汤锅

❷ saucepot 大炖锅

saucepot ['sɔ:spɔt] n.　大炖锅

❸ saucepan 小炖锅

saucepan ['sɔ:spən] n.　小炖锅

❹ frying pan 煎盘

frying ['fraɪɪŋ] v. 油炸，油煎（fry 的现在分词）

pan [pæn] n. 平底锅；盘状的器皿；秤盘

IV Measuring Tools 测量工具

- measuring ['meʒərɪŋ] v. 量（measure 的现在分词）；测量
- tools [tu:lz] n. 工具（tool 的名词复数）

❶ steelyard 秤

steelyard ['sti:ljɑ:d; 'stɪljəd] n. 秤，杆秤

❷ spoon 量勺

spoon [spu:n] n.匙，调羹；一匙的量

❸thermometer 温度计

thermometer[θə'mɒmɪtə]n. 温度计

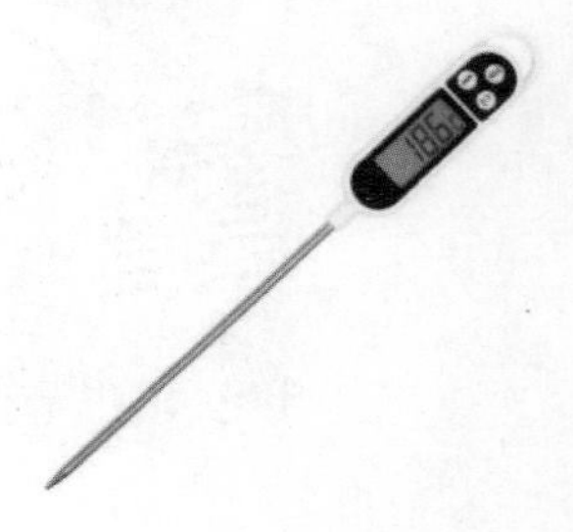

V Knives 刀具

- knife [naɪf] n. 刀；匕首
- knives [naɪvz] n.刀（knife 的名词复数）

❶ chef 's knife 厨师刀

chef [ʃef] n.　厨师，大师傅

❷ boning knife剔骨刀

boning ['bəʊnɪŋ] n. 剔骨，去骨

❸ chopping knife　砍刀

chopping ['tʃɒpɪŋ] n.削球；
[电子] 斩波，斩断

❹ clam knife 蛤刀；牡蛎刀

clam [klæm] n.蚌，蛤

❺ cake knife 锯齿刀；切蛋糕刀

cake [keɪk] n. 蛋糕；糕饼

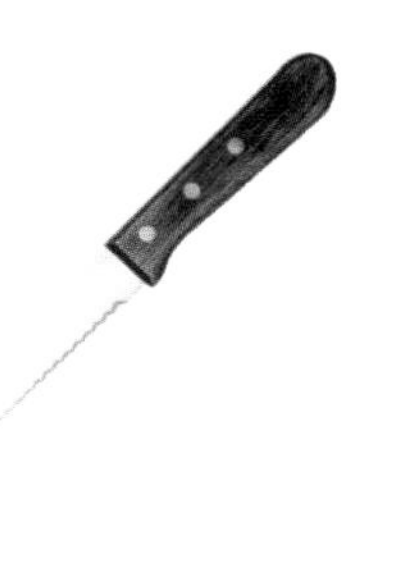

❻ meat pounder 肉锤

meat [mi:t] n.肉；食物
pounder ['paʊndə] n.捣杵；研杵

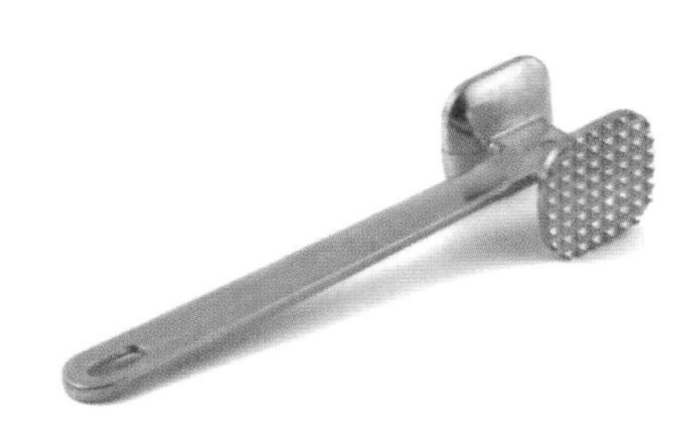

Unit 4

Cooking Methods 烹饪方法

词汇在线

更多烹饪方法

❶ **boil** [bɔɪl] v. 用开水煮

水煮蛋（将食材用开水加工成熟 ）

❷ **grill** [grɪl] v. 烧烤

烤肉（用明火直接烤，尤指户外烧烤）

❸ **braise** [breɪz] v. 炖，焖

炖牛肉（用酱料长时间炖煮大块食材）

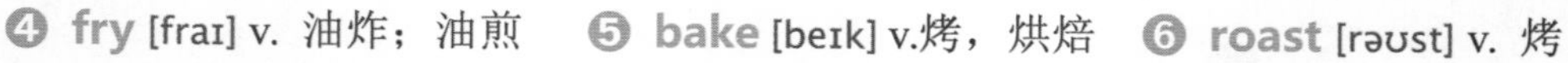

❹ **fry** [fraɪ] v. 油炸；油煎

将食材放入高温热油中加工成熟

❺ **bake** [beɪk] v.烤，烘焙

烤糕点（在密闭的容器中烤，不直接接触火源）

❻ **roast** [rəʊst] v. 烤

烤肉、炒坚果（可在密闭的容器中烤，也可在明火上烤）

❼ **stew** [stjuː] v. 炖

用清水长时间炖煮小块食材

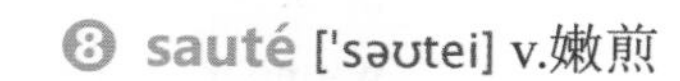

❽ **sauté** ['səʊtei] v.嫩煎

嫩煎，快速成熟

❾ **ice** [aɪs] v. 冰镇

冰镇刺身、啤酒等

Unit 5

Condiments 调味品

词汇在线

❶ salt [sɔ:lt; sɒlt] n. 盐

❷ maple syrup ['meɪp(ə)l 'sɪrəp] n.枫糖浆

❸ honey ['hʌnɪ] n. 蜂蜜

❹ jam [dʒæm] n. 果酱

❺raspberry['rɑ:zb(ə)rɪ] n. 覆盆子；树莓

❻ pungent sauce ['pʌn(d)ʒ(ə)nt sɔ:s]辣酱油

❼ grape wine vinegar [greɪp waɪn 'vɪnɪgə]葡萄酒醋

❽ apple vinegar ['æp(ə)l 'vɪnɪgə]苹果醋

❾ sugar ['ʃʊgə(r)]n.糖

词汇在线

Unit 6

Vanilla and Spice 香草和香料

❶ **rosemary**
['rəʊzm(ə)rɪ] n.迷迭香

❷ **sage**
[seɪdʒ] n.鼠尾草

❸ **basil**
['bæz(ə)l; -zɪl] n.罗勒

❹ **oregano**
[ˌɒrɪ'gɑːnəʊ; ə'regənəʊ] n.
阿里根奴，牛至草，比萨草

❺ **parsley**
['pɑːslɪ] n. 法香，欧芹

❻ **thyme**
[taɪm] n. 百里香

❼ **mint**
[mɪnt] n. 薄荷

❽ **dill**
[dɪl] n. 莳萝

❾ **dill seed**
[dɪl siːd] n. 莳萝子

⑩ cilantro
[sɪ'læntrəʊ] n.芫荽

⑪ black pepper
[blæk 'pepə] 黑胡椒(粉)

⑫ white pepper
[waɪt 'pepə] 白胡椒(粉)

⑬ cinnamon
['sɪnəmən] n. 桂皮；肉桂

⑭ bay-leaf
[beɪ li:f] n.香叶；月桂树叶

⑮ clove
[kləʊv] n.丁香

⑯ cumin
['kʌmɪn] n.小茴香，小茴香子

⑰ mustard
['mʌstəd] n.芥末

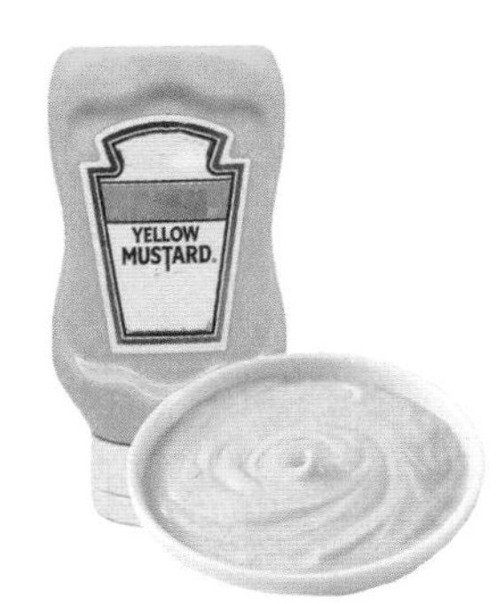

⑱ turmeric
['tɜ:mərɪk] n. 姜黄，姜黄根

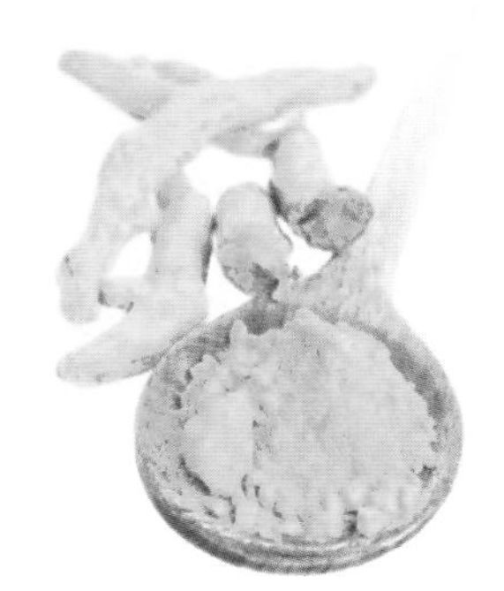

⑲ **horse-radish**

[hɔ:s 'rædɪʃ] n.辣根

⑳ **rhubarb**

['ru:bɑ:b] n. 大黄（茎部）

㉑ **curry powder**

['kʌrɪ 'paʊdə] n. 咖喱粉

㉒ **curry paste**

['kʌrɪ peɪst] n.咖喱酱

㉓ **paprika**

['pæprɪkə;pə'pri:kə]n.红辣椒(粉)

酒类的译法

进口酒类的英文名称仍使用其原文，国产酒类以其注册的英文为准，如果酒类本身没有英文名称，则使用其中文名称的汉语拼音。

Module 2
模块 2

- Unit 7 Sauce Cooking 汁的烹制
- Unit 8 Soup Cooking 汤的烹制
- Unit 9 Vegetable Cooking 蔬菜的烹制
- Unit 10 Potato Cooking 土豆类食物的烹制
- Unit 11 Rice Cooking 米饭类食物的烹制

词汇在线

在线选择

Unit 7

Sauce Cooking 汁的烹制

I Brown Sauce 布朗汁

❶ Learn the following words and phrases.

❷ Match the words and phrases with their meaning.

celery	牛骨头
black pepper	洋葱
onion	肉馅
flour	芹菜
ox bone	胡萝卜
carrot	青蒜
tomato paste	面粉
thyme	番茄膏
minced meat	鲜番茄
green garlic	黑胡椒
potato	香叶
dry red wine	百里香
bay-leaf	土豆
tomato	干红葡萄酒

❸ What's this?

It's____.

It is made of____.

Words List

牛骨头	ox bone [ɔks bəʊn]
肉末，肉馅	minced meat [mɪnst mi:t]
洋葱	onion ['ʌnjən] n.
芹菜	celery ['seləri] n.
胡萝卜	carrot ['kærət] n.
面粉	flour ['flaʊə] n.
黑胡椒	black pepper [blæk 'pepə]
香叶	bay-leaf [bei li:f] n.
百里香	thyme [taim] n.
土豆	potato [pə'teitəʊ] n.
干红	dry red wine [draɪ red wain]
番茄膏	tomato paste [tə'mɑ:təʊ peist]

II Cream Sauce 奶油汁

❶ Learn the following words and phrases.

butter　　milk　　white stock

flour　　pepper

salt　　bay-leaf

❷ Match the words and phrases with their meaning.

salt	黄油
pepper	牛奶
butter	白色基础汤
flour	面粉
milk	盐
bay-leaf	胡椒，胡椒粉
white stock	香叶

❸ What's this?

It's____.

It is made of______________________________________.

Words List

黄油	butter ['bʌtə] n.
牛奶	milk [mɪlk]
清汤；原色汤料	white stock [wait stɔk]
面粉	flour ['flaʊə] n.
盐	salt [sɔ:lt] n.
胡椒；胡椒粉	pepper ['pepə] n.
香叶	bay-leaf [bei li:f] n.

III Hollandaise Sauce 荷兰汁

❶ Learn the following words and phrases.

butter　　yolk　　salt　　lemon juice

❷ Match the words and phrases with their meaning.

yolk	黄油
butter	柠檬汁
lemon juice	盐
salt	鸡蛋黄

❸ What's this?

It's______________________________.

It is made of__________________________.

Words List

黄油	butter ['bʌtə] n.
鸡蛋黄	yolk [jəʊk] n.
盐	salt [sɔːlt] n.
柠檬汁	lemon juice['lemən dʒuːs]

IV Tomato Sauce 番茄汁

❶ Learn the following words and phrases.

tomato paste

basil

bay-leaf

❷ Match the words and phrases with their meaning.

onion	番茄
flour	黄油
butter	香叶
tomato paste	洋葱
bay-leaf	番茄膏
tomato	罗勒
basil	面粉

❸ What's this?

It's____.

It is made of__________________________________.

Words List

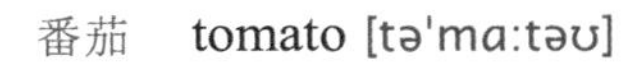

番茄　tomato [tə'mɑ:təʊ]

番茄膏　tomato paste [tə'mɑ:təʊ peist]

罗勒　basil ['bæzəl] n.

洋葱　onion ['ʌnjən] n.

面粉　flour ['flaʊə] n.

黄油　butter ['bʌtə] n.

V Mayonnaise 蛋黄酱

❶ Learn the following words and phrases.

olive oil　mustard　yolk

salt　lemon juice　white pepper

❷ Match the words and phrases with their meaning.

lemon juice	橄榄油
olive oil	蛋黄
yolk	盐
salt	柠檬汁
white pepper	芥末
mustard	白胡椒

❸ What's this?

It's____.

It is made of__________________________________.

Words List

蛋黄酱	Mayonnaise [ˌmeiə'neiz] n.
橄榄油	olive oil ['ɒlɪv ɒɪl]
盐	salt [sɔːlt] n.
柠檬汁	lemon juice ['lemən dʒuːs]
蛋黄	yolk [jəʊk] n.
芥末	mustard ['mʌstəd] n.
白胡椒	white pepper [wait 'pepə]

VI Caesar Dressing 恺撒酱

❶ Learn the following words and phrases.

garlic

Mayonnaise

black pepper

mustard

lemon juice

cheese powder

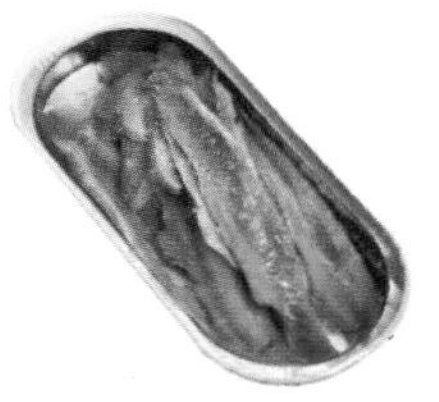

silver fish fillets

❷ Match the words and phrases with their meaning.

black pepper	柠檬汁
garlic	黑胡椒
cheese powder	银鱼柳
silver fish fillets	芥末
lemon juice	奶酪粉
mustard	大蒜

❸ What's this?

It's________________________________.

It is made of____________________________.

Words List

凯撒	Caesar ['si:zə] n.
蛋黄酱	Mayonnaise [,meiə'neiz] n.
大蒜	garlic ['gɑ:lɪk] n.
柠檬汁	lemon juice ['lemən dʒu:s]
芥末	mustard ['mʌstəd] n.
黑胡椒	black pepper [blæk 'pepə]
奶酪粉	cheese powder [tʃi:z 'paʊdə] n.
银鱼	silver fish ['silvə fiʃ] n.
鱼片、鱼柳	fillet ['filit] n.
银鱼柳	silver fish fillets

VII Thousand Island Dressing 千岛酱

❶ Learn the following words and phrases.

Mayonnaise　sour cucumber　garlic　onion

yolk　green pepper　red pepper　tomato sauce

pitted black olive　mustard　lemon juice　chilli powder　black pepper

❷ Match the words and phrases with their meaning.

tomato sauce	洋葱
sour cucumber	大蒜
mustard	辣椒粉
black pepper	酸黄瓜
lemon juice	蛋黄酱
yolk	无核黑橄榄
Mayonnaise	蛋黄
onion	红椒
red pepper	番茄酱汁
pitted black olive	柠檬汁
garlic	黑胡椒
chilli powder	芥末

❸ What's this?

It's____.

It is made of__________________________________.

Words List

一千，一千个	thousand ['θauzənd] n.	黄瓜	cucumber ['kju:kʌmbə] n.
岛，岛屿	island ['ailənd] n.	酸黄瓜	sour cucumber
(拌色拉用)调料；穿衣	dressing['dresɪŋ] n.	大蒜	garlic ['gɑ:lɪk] n.
有凹痕的，去核的	pitted ['pɪtɪd] adj.	洋葱	onion ['ʌnjən] n.
橄榄	olive ['ɒlɪv]	红椒	red pepper [red 'pepə]
无核黑橄榄	pitted black olive	黑胡椒	black pepper [blæk 'pepə]
酸的	sour ['sauə] a.	番茄汁	tomato sauce [tə'mɑ:təʊ sɔ:s]

VIII Vegetable Salad 蔬菜沙拉

❶ Learn the following words and phrases.

onion | red bell pepper | yellow bell pepper

cucumber | iceberg lettuce | radicchio

arugula | cherry tomato | yellow cherry tomato

dill | pitted black olive | thousand island dressing

❷ Match the words and phrases with their meaning.

cherry tomato	黄瓜
iceberg lettuce	洋葱
cucumber	莳萝
onion	红甜椒
radicchio	黄樱桃番茄
arugula	黄甜椒
red bell pepper	红菊苣
yellow cherry tomato	芝麻菜
dill	无核黑橄榄
pitted black olive	红樱桃番茄
thousand island dressing	千岛酱
yellow bell pepper	卷心莴苣球生菜

❸ What's this?

It's____.

It is made of__________________________________.

Words List

红甜椒	red bell pepper [red bel 'pepə]
黄甜椒	yellow bell pepper ['jeləʊ bel 'pepə]
蔬菜沙拉	vegetable salad['vedʒitəbl 'sæləd]
卷心莴苣球(玻璃)生菜	iceberg lettuce ['aisbə:g] ['letis]
红菊苣	radicchio [ræ'di:kjəu] n.
芝麻菜	arugula [æ'ru:gjʊlə] n.
圣女果，红樱桃番茄	cherry tomato ['tʃeri: tə'mɑ:təu]

黄樱桃番茄	yellow cherry tomato
莳萝	dill [dil] n.
有凹痕的，去核的	pitted ['pɪtɪd] adj.
橄榄	olive ['ɒlɪv] n.
无核黑橄榄	pitted black olive
千岛酱	thousand island dressing ['θaʊz(ə)nd 'aɪlənd 'dresɪŋ]

Tips

Parmesan(Italian:Parmigiano-Reggiano),it is named after the producing areas, which comprise the provinces of Parma, Reggio Emilia, Bologna (only the area to the west of the river Reno), Modena (all in Emilia-Romagna), and Mantua (in Lombardy, but only the area to the south of river Po), Italy. Under Italian law, only cheese produced in these provinces may be labelled "Parmigiano-Reggiano", and European law classifies the name, as well as the translation "Parmesan", as a protected designation of origin. Parmigiano is the Italian adjective for Parma and Reggiano is the adjective for Reggio Emilia. Outside the EU, the name "Parmesan" can legally be used for cheeses similar to Parmigiano-Reggiano, with only the full Italian name unambiguously referring to Parmigiano-Reggiano cheese. It has been called the "King of Cheeses".

The king of Cheeses

词汇在线

在线选择

Unit 8

Soup Cooking 汤的烹制

I Soup-Stock 高汤

❶ Learn the following words and phrases.

ox bone

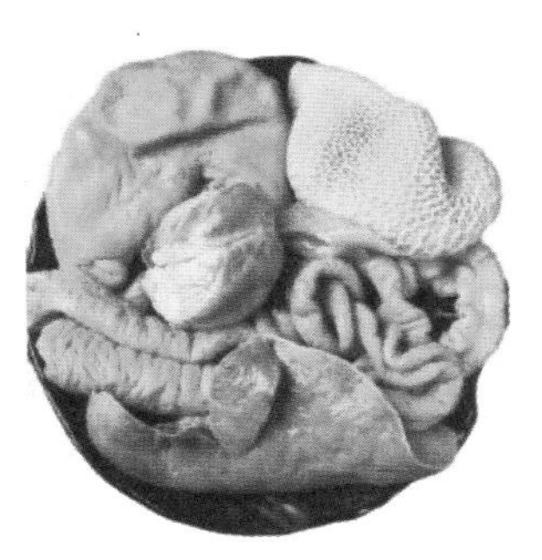

beef offal

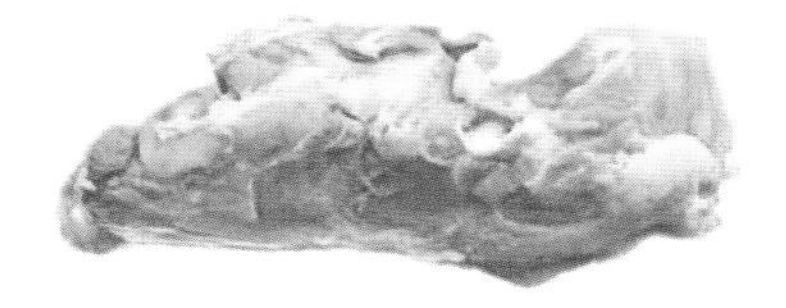

chicken carcass

carrot

celery

❷ Match the words and phrases with their meaning.

beef offal	牛骨
celery	牛杂
carrot	芹菜
chicken carcass	胡萝卜
ox bone	鸡架

❸ What's this?

It's____.

It is made of____.

Words List

牛骨	ox bone [ɔks bəʊn]
牛杂	beef offal [bi:f 'ɔfəl]
（家畜屠宰后的）躯体、骨架等	carcass ['kɑ:kəs] n.
鸡架	chicken carcass ['tʃɪkɪn 'kɑ:kəs]
胡萝卜	carrot ['kærət] n.
芹菜	celery ['selərɪ] n.

II Vegetable Soup 蔬菜汤

❶ Learn the following words and phrases.

red bell pepper

minced garlic

tomato paste

stock

❷ Match the words and phrases with their meaning.

potato	圆白菜
celery	洋葱
tomato paste	橄榄油
red pepper	土豆
olive oil	红椒
cabbage	芹菜
minced garlic	番茄膏
onion	蒜蓉
carrot	高汤
stock	胡萝卜

❸ What's this?

It's____.

It is made of____________________________________.

Words List

芹菜	celery ['seləri] n.
洋葱	onion ['ʌnjən] n.
土豆	potato [pə'teɪtəʊ] n.
胡萝卜	carrot ['kærət] n.
圆白菜	cabbage ['kæbɪdʒ] n.
红辣椒	red pepper [red 'pepə]
高汤	stock [stɒk] n.
橄榄油	olive oil ['ɔliv ɔil]
蒜蓉	minced garlic [mɪnst 'gɑ:lɪk]
番茄膏	tomato paste [tə'mɑ:təu peist]

III Cream Soup 奶油汤

❶ Learn the following words and phrases.

low gluten flour　milk　beef soup

onion　mushroom　bread crumb

butter　cream　chicken essence

bay-leaf　salt　brandy

❷ Match the words and phrases with their meaning.

salt	香叶
low gluten flour	牛肉汤
mushroom	低筋面粉
milk	奶油
beef soup	蘑菇
bay-leaf	洋葱
chicken essence	牛奶
brandy	面包屑
cream	黄油
bread crumb	白兰地
onion	盐
butter	鸡粉；鸡精

❸ What's this?

It's____.

It is made of____________________________________.

Words List

低筋面粉 low gluten flour [ləʊ 'glu:tn 'flaʊə]

牛奶 milk [mɪlk] n.

牛肉汤 beef soup [bi:f su:p]

洋葱 onion ['ʌnjən] n.

蘑菇 mushroom ['mʌʃrʊm] n.

面包屑 bread crumb [bred krʌm]

黄油 butter ['bʌtə] n.

奶油 cream [kri:m] n.

鸡粉 chicken essence ['tʃikin 'esns]

香叶 bay-leaf [bei li:f] n.

盐 salt [sɔ:lt] n.

白兰地 brandy ['brændɪ] n.

IV Clear Soup 清汤

❶ Learn the following words and phrases.

minced beef　beef stock　onion

celery　carrot　garlic

black pepper　bread crumb　bay-leaf

❷ Match the words and phrases with their meaning.

beef stock	黑胡椒
garlic	牛肉高汤
carrot	面包屑
onion	大蒜
minced beef	胡萝卜
bread crumb	芹菜
bay-leaf	牛肉末
black pepper	香叶
celery	洋葱

③ What's this?

It's____.

It is made of____________________________________.

Words List

牛肉末	minced beef [mɪnst bi:f]
牛肉高汤	beef stock [bi:f stɒk]
洋葱	onion ['ʌnjən] n.
芹菜	celery ['selərɪ] n.
胡萝卜	carrot ['kærət] n.
大蒜	garlic ['gɑ:lɪk] n.
黑胡椒	black pepper [blæk 'pepə]
面包屑	bread crumb [bred krʌm]
香叶	bay-leaf [bei li:f] n.

V Seafood Soup 海鲜汤

① Learn the following words and phrases.

prawn

mussel

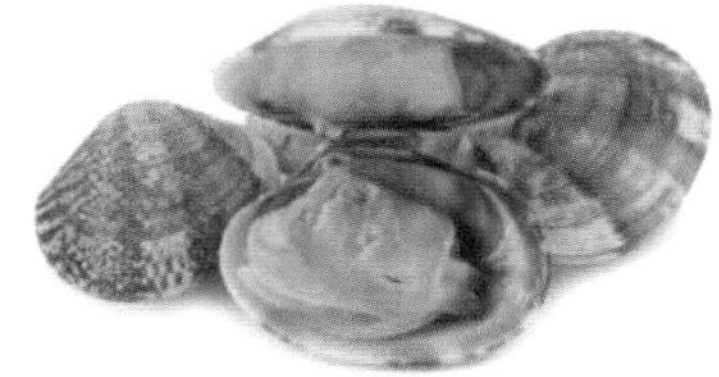

clam

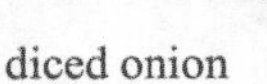

diced onion　　silvery pout　　tomato paste　　chilli powder

❷ Match the words and phrases with their meaning.

diced onion	蛤蜊
prawn	洋葱粒
mussel	辣椒粉
clam	贻贝
chili powder	对虾
silvery pout	番茄膏
tomato paste	银鳕鱼

❸ What's this?

It's____.

It is made of___________________________________.

Words List

对虾，明虾　prawn [prɔ:n] n.

青口，贻贝　mussel ['mʌsl] n.

蛤蜊　clam [klæm] n.

洋葱粒　diced onion [daist 'ʌnjən]

辣椒粉　chilli powder ['tʃɪlɪ 'paʊdə]

银鳕鱼　silvery pout ['sɪlvəri: paʊt]

番茄膏　tomato paste [tə'mɑ:təu peist]

词汇在线

Unit 9

Vegetable Cooking 蔬菜的烹制

在线选择

I Stewed Pearl Onion 红烩珍珠洋葱

❶ Learn the following words and phrases.

butter　　white pepper　　brown sauce

pearl onion　　salt　　red wine

❷ Match the words and phrases with their meaning.

red wine	炖；煨
pearl onion	黄油
butter	红酒
brown sauce	白胡椒
salt	珍珠洋葱
stewed	布朗汁
white pepper	盐

③ What's this?

It's____.

It is made of________________________________.

Words List

炖（stew 的过去式和过去分词）；煨	stewed [stju:d] v.
珍珠洋葱	pearl onion [pɜ:l 'ʌnjən]
盐	salt [sɔ:lt] n.
黄油	butter ['bʌtɚ] n.
布朗汁，上色汁	brown sauce [braʊn sɔ:s]
红色的	red [red] adj.
葡萄酒；果酒	wine [waɪn] n.
红酒；红葡萄酒	red wine [red waɪn]
白胡椒	white pepper [waɪt 'pepə]

II Fried Egg-plant with Tomato Sauce 煎茄子配番茄汁

① Learn the following words and phrases.

tomato sauce

egg-plant

white pepper

❷ Match the words and phrases with their meaning.

bread crumb	面粉
fried	盐
egg-plant	面包屑
tomato sauce	油炸的，油煎的
salt	油
flour	番茄酱汁
oil	茄子

❸ What's this?

It's____.

It is made of__________________________________.

Words List

油炸的，油煎的	fried [fraɪd] adj.
茄子	egg-plant [eg plɑ:nt] n.
油	oil [ɔɪl] n.
盐	salt [sɔ:lt] n.
番茄酱汁	tomato sauce [tə'mɑ:təʊ sɔ:s]
番茄	tomato [tə'mɑ:təʊ] n.
调味汁；酱汁	sauce [sɔ:s] n.
面粉	flour ['flaʊə] n.
蛋清	egg white [eg waɪt]
碎屑（尤指面包屑或糕饼屑）	crumb [krʌm] n.
面包屑	bread crumb [bred krʌm]
白胡椒	white pepper [waɪt 'pepə]

Ⅲ Provence Fried Cucumber 普罗旺斯炒胡瓜

❶ Learn the following words and phrases.

❷ Match the words and phrases with their meaning.

zucchini	橄榄油
Provence	盐
onion	节瓜
cucumber	洋葱（头）
salt	大蒜
garlic	普罗旺斯
olive oil	黄瓜

❸ What's this?

It's____.

It is made of__________________________________.

Words List

普罗旺斯（法国东南部一地区）	**Provence** [prɔ'vaːns] n.
黄瓜	**cucumber** ['kjuːkʌmbə] n.
节瓜	**zucchini** [zʊ'kiːnɪ] n.
橄榄油	**olive oil** ['ɒlɪv ɒɪl]
橄榄	**olive** ['ɒlɪv] n.
大蒜；蒜头	**garlic** ['gaːlik] n.

IV Roasted Garlic 烤蒜

❶ Learn the following words and phrases.

garlic

olive oil

salt

black pepper

❷ Match the words and phrases with their meaning.

olive oil	大蒜
roasted	黑胡椒
garlic	盐
salt	橄榄油
black pepper	烤，烘，焙

❸ What's this?

It's____.

It is made of____________________________________.

Words List

烤，烘，焙（roast 的过去式和过去分词）	roasted ['rəʊstɪd]
烤，烘，焙	roast ['rəʊst] v.
大蒜	garlic ['gɑːlɪk] n.
盐	salt [sɔːlt] n.
黑色	black [blæk] n.
胡椒；辣椒；胡椒粉	pepper ['pepə] n.
橄榄油	olive oil ['ɒlɪv ɒɪl]

V Baked Sliced Tomato 烤番茄片

❶ Learn the following words and phrases.

tomato butter salt white pepper

❷ Match the words and phrases with their meaning.

baked	刨切的
butter	白胡椒
sliced	黄油
tomato	盐
salt	烘烤制作的
white pepper	番茄

❸ What's this?

It's________________________.

It is made of________________________.

Words List

烘烤制作的	baked [beɪkt] adj.
刨切的	sliced [slaist] adj.
番茄	tomato [tə'mɑ:təʊ] n.
盐	salt [sɔ:lt] n.
白胡椒	white pepper [waɪt 'pepə]
黄油	butter ['bʌtə] n.

VI Stuffed Tomato with Millet 粟米酿番茄

❶ Learn the following words and phrases.

❷ Match the words and phrases with their meaning.

stuffed	番茄
shallot	鲜奶油
sweet corn	把……装满
tomato	黄油
butter	干葱头
whipping cream	甜玉米

❸ What's this?

It's____.

It is made of____________________________________.

Words List

塞住（stuff 的过去式和过去分词）；把……装进；把……装满	stuffed [stʌft] v.
番茄	tomato [tə'mɑ:təʊ] n.
小米；粟；稷	millet ['mɪlɪt] n.
甜玉米；粟米	sweet corn[swi:t kɔ:n]
鞭打（whip 的现在分词）；抽打；搅拌……至变稠	whipping ['wɪpɪŋ] n.
（乳脂含量较高的）鲜奶油	whipping cream ['wɪpɪŋ kri:m]
干葱头	shallot [ʃə'lɒt] n.

VII Italian Seasonable Vegetables Rolls 意大利蔬菜卷

❶ Learn the following words and phrases.

❷ Match the words and phrases with their meaning.

cucumber	蔬菜
seasonable	黄椒
red leaf lettuce	卷
vegetable	合时令的
rolls	黄瓜
yellow pepper	红叶生菜

❸ What's this?

It's____.

It is made of__________________________________.

Words List

意大利的	Italian [ɪ'tæljən] adj.
合时令的	seasonable ['si:znəbl] adj.
蔬菜	vegetable ['vedʒtəbl] n.
滚翻（roll 的名词复数）；卷	rolls [rɒlz] n.
红椒	red pepper [red 'pepə]
黄椒	yellow pepper ['jeləʊ 'pepə]
西葫芦	squash [skwɒʃ] n.
红叶生菜	red leaf lettuce [red li:f 'letɪs]
叶子	leaf [li:f] n.
莴苣，生菜	lettuce ['letɪs] n.

VIII Fresh Asparagus with Hollandaise Sauce 荷兰汁鲜芦笋

❶ Learn the following words and phrases.

asparagus

egg

lemon

salt

white pepper

pure butter

❷ Match the words and phrases with their meaning.

asparagus	新鲜的
lemon	荷兰汁
fresh	芦笋
pure butter	柠檬
Hollandaise Sauce	清黄油

❸ What's this?

It's____.

It is made of____________________________________.

Words List

新鲜的	fresh [freʃ] adj.
芦笋	asparagus [ə'spærəgəs] n.
荷兰	Holland ['hɒlənd] n.
果汁；肉汁	juice [dʒu:s] n.
清黄油	pure butter [pjʊə 'bʌtə]
纯的；单纯的；干净的	pure [pjʊə] adj.
柠檬	lemon ['lemən] n.

词汇在线

Unit 10

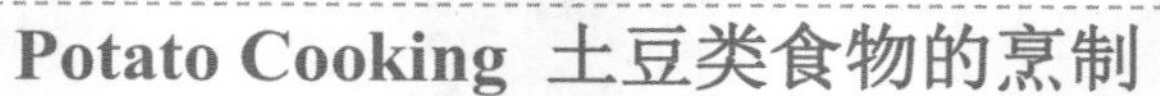

Potato Cooking 土豆类食物的烹制

在线选择

I Roast Potato on Tinfoil Pork 锡纸烤土豆

❶ Learn the following words and phrases.

potato　bacon

meat floss　thousand island dressing　chives

❷ Match the words and phrases with their meaning.

meat floss	培根
potato	细香葱
thousand island dressing	肉松
chives	土豆
bacon	千岛酱

③ What's this?

It's____.

It is made of________________________________.

Words List

土豆	potato [pə'teɪtəʊ] n.
培根	bacon ['beɪk(ə)n] n.
葱	chives [tʃaɪvz] n.
肉松	meat floss [mi:t flɒs]
千岛酱	thousand island dressing ['θaʊz(ə)nd 'aɪlənd 'dresɪŋ]

Ⅱ Fried Potato Balls 炸薯球

① Learn the following words and phrases.

potato onion egg

flour crumb minced beef

pea

white pepper

oil

❷ Match the words and phrases with their meaning.

onion	白胡椒
crumb	面包屑
white pepper	油
oil	洋葱
flour	面粉

❸ What's this?

It's____.

It is made of______________________________________.

Words List

土豆　potato [pəˈteɪtəʊ] n.
洋葱　onion [ˈʌnjən] n.
面粉　flour [flaʊə] n.
鸡蛋　egg [eg] n.
盐　salt [sɔːlt] n.

III Roasted Dices Potato with Rosemary 迷迭香烤薯角

❶ Learn the following words and phrases.

❷ Match the words and phrases with their meaning.

roast	烤
rosemary	土豆
salt	迷迭香
potato	橄榄油
olive oil	盐

❸ What's this?

It's____.

It is made of____________________________________.

Words List

迷迭香	rosemary ['rəʊzm(ə)rɪ] n.
烤	roast [rəʊst] v.
盐	salt [sɔːlt] n.
橄榄油	olive oil ['ɔliv ɔil]

IV Fried Golden Potato Biscuits 煎黄金薯饼

❶ Learn the following words and phrases.

potato

shredded potato

butter

salt

❷ Match the words and phrases with their meaning.

grate	土豆丝
peel	土豆
potato	去皮
butter	刨丝
shredded potato	黄油

❸ What's this?

It's________________________________.

It is made of________________________________.

Words List

土豆丝	shredded potato [ʃredid pə'teɪtəʊ]
去皮	peel [pi:l] v.
刨丝	grate [greɪt] v.
煎	fry [fraɪ] v.

V Mashed Potato with Egg 蛋黄土豆泥

❶ Learn the following words and phrases.

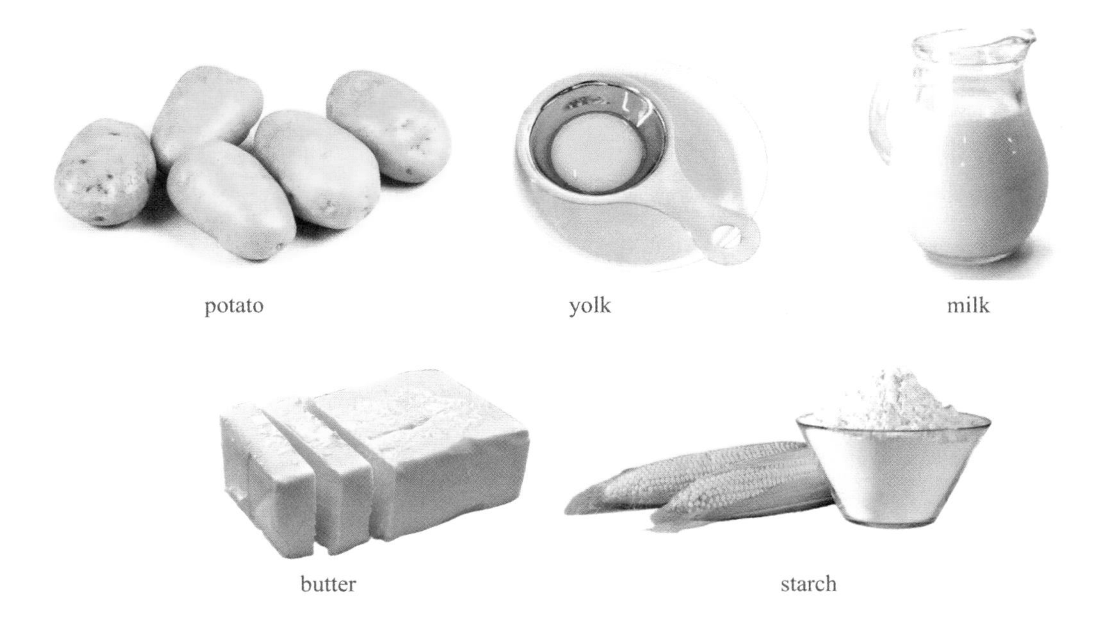
potato yolk milk
butter starch

❷ Match the words and phrases with their meaning.

yolk	黄油
milk	鸡蛋黄
butter	土豆
salt	牛奶
potato	芡粉
starch	盐

❸ What's this?

It's____.

It is made of______________________________________.

Words List

鸡蛋黄	yolk [jəʊk] n.
牛奶	milk [mɪlk] n.
芡粉	starch [stɑːtʃ] n.

VI Baked Diced Potato with Butter 黄油焗薯块

❶ Learn the following words and phrases.

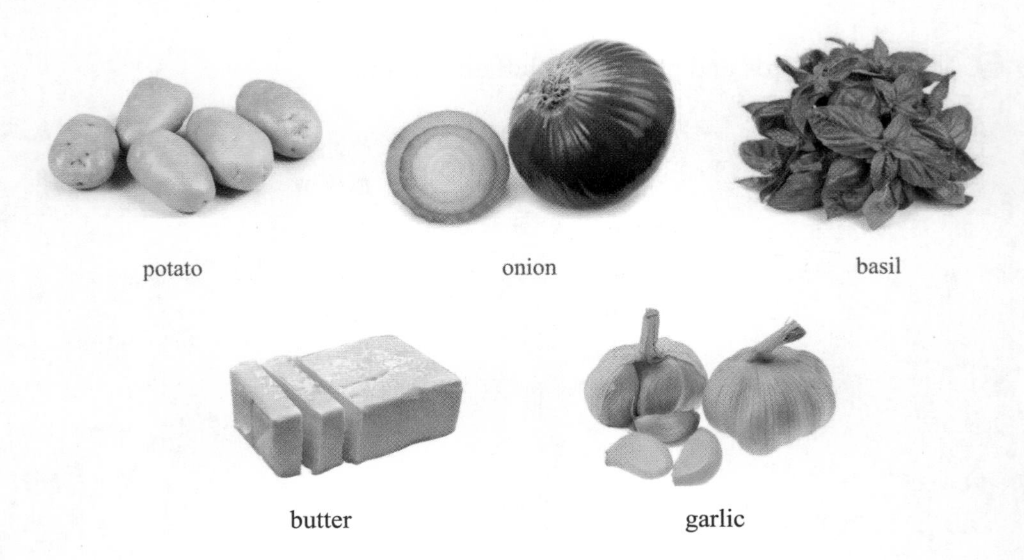

potato　　onion　　basil

butter　　garlic

❷ Match the words and phrases with their meaning.

butter	黄油
onion	土豆
basil	盐
salt	大蒜
potato	洋葱
garlic	罗勒

❸ What's this?

It's______________________________.

It is made of__________________________.

Words List

罗勒	basil ['bæzl] n.
洋葱	onion ['ʌnjən] n.
大蒜	garlic ['gɑːlɪk] n.
黑胡椒	black pepper [blæk 'pepə]

VII Fried Chips 炸薯条

❶ Learn the following words and phrases.

potato pepper salt ketchup

❷ Match the words and phrases with their meaning.

potato chip	胡椒
fry	土豆条
pepper	土豆
salt	盐
Potato	沥干
drain	油炸

❸ What's this?

It's____.

It is made of__________________________________.

Words List

土豆	potato [pə'teɪtəʊ] n.
（食物的）小片；薄片	chip [tʃɪp] n.
盐	salt [sɔ:lt] n.
胡椒	pepper ['pepə] n.
油炸	deep-fry ['di:p'frai] v.
沥干	drain [dreɪn] v.
番茄酱（加了调味料，多蘸薯条吃）	ketchup ['ketʃəp] n.

Tips

Balsamic vinegar (Italian: aceto balsamico) is a very dark, concentrated, and intensely flavoured vinegar made wholly or partially from grape must, originating in Italy.The term "aceto balsamico" is unregulated, but there are three protected balsamic vinegars: "Aceto Balsamico Tradizionale di Modena" (Traditional Balsamic Vinegar of Modena), "Aceto Balsamico Tradizionale di Reggio Emilia" (Traditional Balsamic Vinegar of Reggio Emilia), and "Aceto Balsamico di Modena" (Balsamic Vinegar of Modena). The two traditional balsamic vinegars are made the same way from reduced grape must aged for several years in a series of wooden barrels, and are produced exclusively in either the province of Modena or the wider Emilia region surrounding it. The names of these two vinegars are protected by the European Union's Protected Designation of Origin, so they are expensive. while Balsamic Vinegar of Modena (Aceto Balsamico di Modena) is made from grape must blended with wine vinegar,so it is usually not so expensive.

Balsamic vinegar

词汇在线

Unit 11

Rice Cooking 米饭类食物的烹制

在线选择

I Paella 西班牙海鲜饭

❶ Learn the following words and phrases.

rice squid shrimp

clam mussel olive oil

rosemary saffron white wine black pepper

red pepper lemon onion

❷ Match the words and phrases with their meaning.

rice	虾
squid	藏红花
shrimp	贻贝
clam	蛤蜊
mussels	大米
Saffron	迷迭香
rosemary	鱿鱼

❸ What's this?

It's____.

It is made of____________________________________.

❹ Classify the following words or phrases.

squid shrimp clam mussel saffron
rosemary black pepper red pepper
onion red pepper potato

Seafood（海鲜）	Spices（香料）	Vegetable（蔬菜）
____________	____________	____________
____________	____________	____________
____________	____________	____________

Words List

西班牙海鲜饭	paella [paɪ'elə] n.
稻；稻米，大米	rice [raɪs] n.
鱿鱼	squid [skwɪd] n.
虾	shrimp [ʃrɪmp] n.
蛤蜊	clam [klæm] n.
青口，贻贝	mussel ['mʌsl] n.
[植] 藏红花，番红花	saffron ['sæfrən] n.

II Butter Rice 黄油米饭

❶ Learn the following words and phrases.

leftover rice　　butter　　egg

sugar　　white sesame

❷ Match the words and phrases with their meaning.

leftover rice	白砂糖
butter	剩米饭
white sesame	黄油
egg	白芝麻
sugar	鸡蛋

③ What's this?

It's____.

It is made of__________________________________.

Words List

剩余的；未用完的；吃剩的	leftover ['leftəʊvə] adj.
剩饭	leftover rice ['leftəʊvə raɪs]
芝麻	sesame ['sesəmɪ] n.
白芝麻	white sesame [waɪt 'sesəmɪ]

III Saffron Rice 红花饭

① Learn the following words and phrases.

rice

saffron

chive

chicken soup

turmeric powder

salt

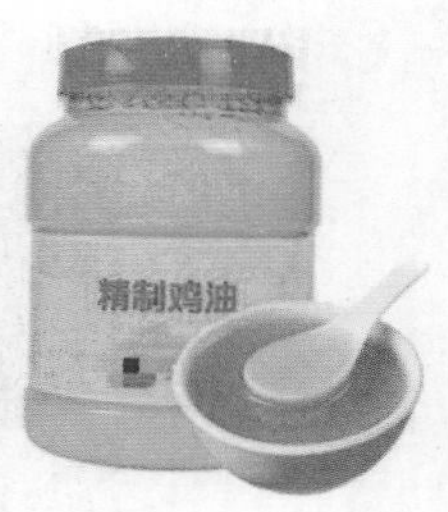

chicken oil

❷ Match the words and phrases with their meaning.

saffron	鸡汤
chicken oil	味精
chicken soup	姜黄粉
turmeric powder	鸡油
monosodium glutamate	藏红花

❸ What's this?

It's____________________________________.

It is made of______________________________.

Words List

鸡油	chicken oil ['tʃɪkɪn ɒɪl]
鸡汤	chicken soup ['tʃɪkɪn su:p]
姜黄，姜黄根	turmeric ['tɜ:mərɪk] n
姜黄粉	turmeric powder ['tɜ:mərɪk 'paʊdə]

IV Curry Fried Rice 咖喱炒饭

❶ Learn the following words and phrases.

onion　tomato　green pepper

egg　curry powder　vegetarian ham

asparagus　parsley

olive oil　steamed rice

❷ Match the words and phrases with their meaning.

asparagus	欧芹
vegetarian ham	咖喱粉
curry powder	橄榄油
olive oil	芦笋
parsley	素火腿

❸ What's this?

It's____.

It is made of__________________________________.

Words List

芦笋，芦笋的茎 asparagus [ə'spærəgəs] n.

素食者的；素菜的 vegetarian [ˌvedʒɪ'teərɪən] adj.

素火腿 vegetarian ham [ˌvedʒɪ'teərɪən hæm]

Tips 中文菜名英文译法 ❶

以主料为主，配料或配汁为辅

类别	中文菜名示例	英文译法示例	翻译原则
主料＋配料	松仁香菇	Chinese Mushrooms with Pine Nuts	主料(名称/形状)+with+配料
主料＋配汁	冰梅凉瓜	Bitter Melon in Plum Sauce	主料+with/in+汤汁(Sauce)

Module 模块 3

3

词汇在线

Unit 12

Beef Cooking 牛肉的烹制

在线选择

I Grilled Garlic Sirloin Steak 扒蒜蓉西冷牛排

❶ Learn the following words and phrases.

sirloin chips butter

chopped garlic broccoli tomato

white pepper oil salt

❷ Match the words and phrases with their meaning.

sirloin	西兰花
chips	白胡椒
butter	油
garlic	盐
white pepper	牛里脊肉
broccoli	炸马铃薯条
tomato	番茄
oil	黄油
salt	大蒜

❸ What's this?

It's____.

It is made of____.

Words List

牛里脊肉	**sirloin** ['sɜ:lɒɪn] **n.**
炸马铃薯条	**chips** [tʃips] **n.**
黄油	**butter** ['bʌtə] **n.**
大蒜	**garlic** ['ɡɑ:lik] **n.**
剁碎的	**chopped** [tʃɒpt] **adj.**
蒜末	**chopped garlic** [tʃɒpt 'ɡɑ:lik]
胡椒	**pepper** ['pepə] **n.**
白胡椒	**white pepper** [waɪt 'pepə]
花椰菜；西兰花	**broccoli** ['brɒkəli] **n.**
番茄	**tomato** [tə'mɑ:təʊ] **n.**
油	**oil** [ɒɪl] **n.**
盐	**salt** [sɔ:lt] **n**

Ⅱ Roasted Beef Fillet with Black Pepper 黑椒烤牛柳

❶ Learn the following words and phrases.

❷ Match the words and phrases with their meaning.

A

beef fillet	黄瓜
onion	小番茄
chives	牛柳
cucumber	欧芹
small tomato	细香葱
parsley	洋葱

B

thyme	黑胡椒汁
rosemary	布朗汁
meat tenderizer	蛋黄酱
black pepper sauce	迷迭香
brown sauce	白兰地
brandy	松肉粉
chopped garlic	蒜末
mayonnaise	百里香

❸ What's this?

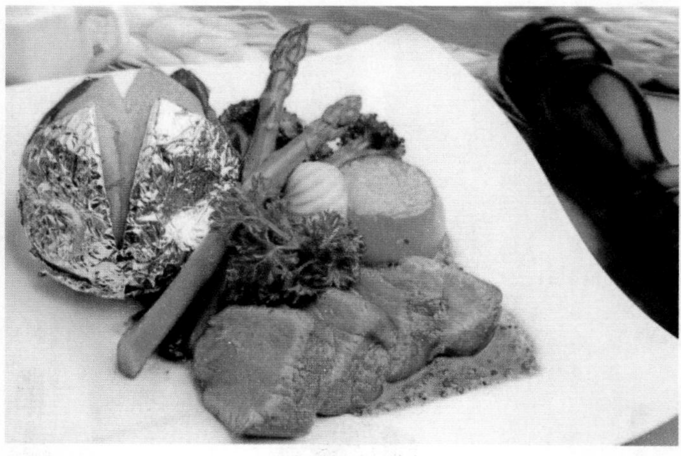

It's________________________________.

It is made of________________________________.

Words List

肉片，鱼片；把（鱼、肉）切成片	fillet ['fɪlɪt] n. & vt..
牛肉	beef [bi:f] n.
牛柳	beef fillet [bi:f 'fɪlɪt]
洋葱	onion ['ʌnjən] n.
细香葱	chives [tʃaɪvz] n.
黄瓜	cucumber ['kju:kʌmbə] n.
法香，欧芹	parsley ['pɑ:slɪ] n.
百里香	thyme [taɪm] n.
迷迭香	rosemary ['rəʊzm(ə)rɪ] n.
肉	meat [mi:t] n.
（肉类）嫩化剂；嫩肉粉	tenderizer ['tendəraɪzə] n.
松肉粉	meat tenderizer [mi:t 'tendəraɪzə]
黑色	black [blæk] n.
黑胡椒	black pepper [blæk 'pepə]
汁，沙司，酱	sauce [sɔ:s] n.
黑胡椒汁	black pepper sauce [blæk 'pepə sɔ:s]

III T-bone Steak with Shallot and Brandy Sauce T骨牛排配干葱白兰地汁

❶ Learn the following words and phrases.

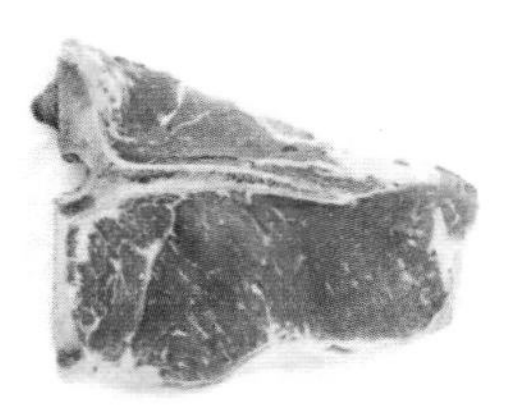

T-bone steak

zucchini

bell pepper

potato

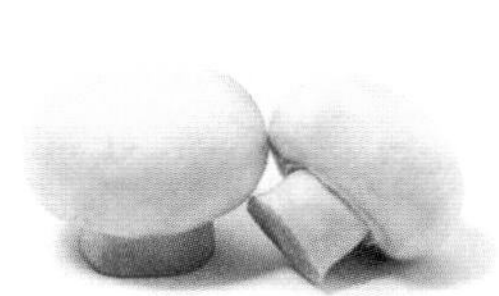

mushroom

shallot bulb

❷ Match the words and phrases with their meaning.

T-bone steak	土豆
zucchini	蘑菇
bell pepper	红酒
potato	节瓜
mushroom	T 骨牛排
shallot bulb	柿子椒
red wine	胡葱

❸ What's this?

It's____.

It is made of__________________________________.

Words List

骨头	bone [bəʊn] n.
牛排；肉排；鱼排	steak [steɪk] n.
T 骨牛排	T-bone steak ['ti:bəʊn steɪk]
节瓜	zucchini [zʊ'ki:nɪ] n.
钟，铃；钟声，铃声；钟状物	bell [bel] n.
菜椒；柿子椒	bell pepper [bel 'pepə]
马铃薯，土豆	potato [pə'teɪtəʊ] n.
蘑菇；蘑菇状物	mushroom ['mʌʃru:m; -rʊm] n.
葱；红葱；青葱	shallot [ʃə'lɒt] n.
球茎，块茎植物；电灯泡	bulb [bʌlb] n.
胡葱	shallot bulb [ʃə'lɒt bʌlb]
红色的	red [red] adj.
葡萄酒；果酒	wine [waɪn] n.
红酒	red wine [red waɪn]

IV Grilled Beef Tenderloin with Curry Sauce 扒牛柳咖喱汁

❶ Learn the following words and phrases.

beef tenderloin

green pepper

red bell pepper

broccoli

potato

thyme

coconut milk

curry

banana puree

❷ Match the words and phrases with their meaning.

green pepper	香蕉泥
red pepper	红酒
banana puree	红椒
coconut milk	肉桂粉
curry	青椒
cinnamon	椰奶
red wine	咖喱

❸ What's this?

It's____________________________________.

It is made of________________________________.

④ Classify the following words or phrases.

green pepper	red pepper	broccoli	onion
white pepper	potato	thyme	coconut
curry	banana	cinnamon	red wine

Fruit（水果）	Spices（香料）	Wine（酒）	Vegetable（蔬菜）
________	________	________	________
________	________	________	________
________	________	________	________

Words List

绿色的	green [gri:n] adj.
青椒	green pepper [gri:n 'pepə]
红椒	red pepper [red 'pepə]
[植] 椰子；[植] 椰肉	coconut ['kəʊkənʌt] n.
椰奶	coconut milk ['kəʊkənʌt mɪlk]
咖喱	curry ['kʌrɪ] n.
香蕉	banana [bə'nɑ:nə] n.
浓汤；果泥；菜泥	puree ['pjʊəreɪ] n.
香蕉泥	banana puree [bə'nɑ:nə 'pu:rɪ]
肉桂；桂皮香料	cinnamon ['sɪnəmən] n.

V Veal Cordon Blue 歌顿堡牛仔排

① Learn the following words and phrases.

beef fillet

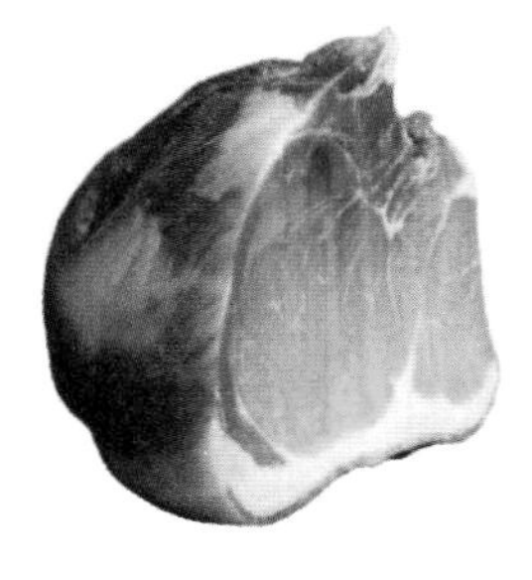

parma ham

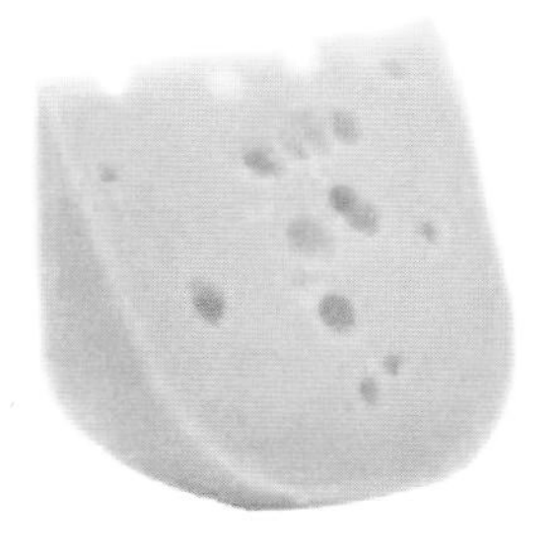

cheese

egg

flour

bread crumb

lea perrins

balsamico

red wine

vegetable seeds oil

olive oil

squash

red bell pepper

yellow bell pepper

eggplant

salt

white pepper

❷ Match the words and phrases with their meaning.

white pepper	白胡椒
parma ham	意大利黑醋
lea perrins	西葫芦
squash	茄子
eggplant	黄椒
yellow pepper	李派林
balsamico	帕尔玛火腿

❸ What's this?

It's__________________________________.

It is made of______________________________.

❹ Classify the following words or phrases.

sirloin parma ham lea perrins
balsamico squash eggplant red pepper
yellow pepper salt oil olive oil

Meat（肉）	Condiment（调味品）	Oil（油）	Vegetable（蔬菜）
________	________	________	________
________	________	________	________
________	________	________	________

Words List

牛肉；小牛	veal [vi:l] n.
[法] 蓝绶带，蓝绶带骑士	Cordon Blue ['kɔɪd(ə)n blu:]
帕尔玛（意大利北部城市）	Parma ['pɑ:mə] n.
火腿	ham [hæm] n.
帕尔玛火腿	Parma ham ['pɑ:mə hæm]
李派林；辣酱油	lea perrins [li: 'perinz]
笋瓜；西葫芦	squash [skwɒʃ] n.
茄子	eggplant ['egplɑ:nt] n.
意大利黑醋	balsamico [bɒl'sæmɪkə] n.
白色的	white [waɪt] adj.
白胡椒	white pepper [waɪt 'pepə]
种子	seed [si:d] n.
菜籽油	vegetable seeds oil ['vedʒtəb(ə)l si:dz ɒɪl]

Tips

Parma ham(Prosciuttodi Parma), the origin is southern mountains in Parma, Italy. Parma ham, the world's most famous prosciutto, its colour and lustre is like a pink rose, uniform fat distribution, the most soft mouthfeel in various ham, therefore,authentic Italian restaurants all supply it. In Italy, the availability of quality Parma ham is almost a standard for evaluating the quality of restaurants.

VI Veal Roll with Spinach 小牛肉菠菜卷

❶ Learn the following words and phrases.

veal fillet spinach taro

brown sauce white pepper mustard

lea perrins salt whipping cream red wine

❷ Match the words and phrases with their meaning.

taro	李派林
mustard	小牛肉
whipping cream	芥末
lea perrins	鲜奶油
veal	香芋

❸ What's this?

It's____.

It is made of_______________________________________.

Words List

牛肉；小牛 veal [vi:l] n.

小牛仔肉 veal fillet [vi:l 'fɪlɪt]

卷饼；小圆面包；牛肠 roll [rəʊl] n.

芋头 taro ['tɑ:rəʊ; tærəʊ] n.

鞭打（whip 的现在分词）；抽打；搅拌……直至变稠 whipping ['wɪpɪŋ] v.

（乳脂含量较高的）鲜奶油 whipping cream ['wɪpɪŋ kri:m]

中文菜名英文译法 ❷

原料可数名词单复数使用

类别	中文菜名示例	英文译法示例	翻译原则
可数名词原料	蔬菜面	Noodles with Vegetables	(1) 原料是可数名词的，基本用复数 (2) 在整道菜中只用了一件原料，或原料太细碎无法数清时，用单数
	葱爆羊肉	Sautéed Lamb Slices with Scallion	

词汇在线

Unit 13

Pork Cooking 猪肉的烹制

在线选择

I Baked Pork Chops with Prune Stuffing 烤酿馅猪排

❶ Learn the following words and phrases.

pork chop

prune

white wine

sherry

brown sauce

white pepper

❷ Match the words and phrases with their meaning.

stuff	雪利酒
bake	盐
pork chop	西梅干
sherry	猪排
salt	白胡椒
white pepper	白葡萄酒
white wine	布朗汁
brown sauce	烤
prune	填塞

❸ What's this?

It's____.

It is made of_______________________________________.

Words List

雪利酒	sherry ['ʃerɪ; 'ʃɛrɪ] n.
西梅干	prune [pru:n] n.
塞满，填塞	stuff [stʌf] vt.
烤，烘焙	bake [beɪk] vt. & vi.
猪肉	pork [pɔ:k] n.
猪排	pork chop [pɔ:k tʃɒp]
布朗汁（少司）	brown sauce [braʊn sɔ:s]

II Grilled Marinated Pork Tenderloin with Sweet Potato Puree and Warm Chipotle 架烤猪里脊配甘薯泥、墨西哥热番茄辣酱

❶ Learn the following words and phrases.

❷ Match the words and phrases with their meaning.

A

grilled	甘薯
marinated	果酱，菜泥
pork tenderloin	温暖的
sweet potato	腌，浸泡
puree	墨西哥番茄辣酱
warm	猪里脊
chipotle salsa	架烤的

B

oregano	橄榄油
cumin powder	青柠檬汁
cinnamon	红椒
olive oil	牛至叶
small tomato	薄荷
green lemon juice	孜然粉
garlic	小番茄
red pepper	辣椒酱
chilli sauce	大蒜
mint	肉桂

❸ What's this?

It's____.

It is made of____________________________________.

❹ Classify the following words or phrases.

sweet potato	chipotle salsa	oregano	cumin powder
mint	small tomato	onion	garlic
red pepper	chilli sauce	cinnamon	

Sauce（调味汁）	Spices（香料）	Vegetable（蔬菜）
____________	____________	____________
____________	____________	____________
____________	____________	____________

Words List

用腌汁浸泡；腌泡	marinate ['mærɪneɪt] vt. & vi.
（牛、羊等的）腰部嫩肉；里脊	tenderloin ['tendəlɒɪn] n.
猪里脊	pork tenderloin [pɔ:k 'tendəlɒɪn]
甘薯	sweet potato [swi:t pə'teɪtəʊ]
暖和的，温暖的	warm [wɔ:m] adj.
干红辣椒，烟味辣椒，墨西哥风味 快速便餐	chipotle [tʃɪ'pəʊtlɪ] n.
洋葱做的辣调味汁，辛香番茄酱（吃墨西哥或西班牙食物时用）	salsa ['sælsə] n.
比萨草，牛至草，阿里根奴	oregano [ˌɒrɪ'gɑ:nəʊ; ə'regənəʊ] n.
小茴香；莳萝	cumin ['kʌmɪn] n.
粉，粉末；粉状物质	powder ['paʊdə] n.
樟属植物；桂皮香料；肉桂色	cinnamon ['sɪnəmən] n.
辣椒酱	chilli sauce ['tʃili sɔ:s]
薄荷	mint [mɪnt] n.
浓汤；果泥；菜泥	puree ['pu:rɪ] n.
绿色的；未熟的	green [gri:n] adj.
柠檬	lemon ['lemən] n.
果汁；肉汁	juice [dʒu:s] n.
青柠檬汁	green lemon juice [gri:n 'lemən dʒu:s]

III Pork Roll and Yellow Peach with Mushroom Sauce 黄桃猪肉卷配白菌汁

❶ Learn the following words and phrases.

pork loin

yellow peach

onion

spinach

mustard

potato

mushroom

❷ Match the words and phrases with their meaning.

pork loin	白菌
yellow peach	菠菜
mustard	猪外脊
spinach	芥末
mushroom	黄桃

③ What's this?

It's____.

It is made of______________________________________.

Words List

腰部；（牛、羊等的）腰肉	loin [lɒɪn] n.
猪大排骨，猪外脊	pork loin [pɔ:k lɒɪn]
桃子	peach [pi:tʃ] n.
黄桃	yellow peach ['jeləʊ pi:tʃ]
芥末	mustard ['mʌstəd] n.
菠菜	spinach ['spɪnɪdʒ; -ɪtʃ] n.

IV Bacon Wrapped Pork Tenderloin 扒腌肉卷猪柳

❶ Learn the following words and phrases.

pork tenderloin

bacon

potato　　parsley　　broccoli

black pepper

carrot

❷ Match the words and phrases with their meaning.

bacon	土豆
wrapped	西兰花
potato	黑胡椒
black pepper	胡萝卜
parsley	番芫荽
broccoli	腌肉
carrot	有包装的

❸ What's this?

It's____.

It is made of__________________________________.

Words List

培根，腌肉	bacon ['beɪk(ə)n] n.
包，缠绕，用……包裹	wrap [ræp] vt.
有包装的，预先包装的	wrapped [ræpt] adj.
黑胡椒	black pepper [blæk 'pepə]
法香，番芫荽，欧芹，荷兰芹	parsley ['pɑːslɪ] n.
胡萝卜	carrot ['kærət] n.

V Italian Pork Chop with Tomato Sauce 茄汁意大利猪排

❶ Learn the following words and phrases.

pork loin　egg　flour

red wine　white pepper　salt

spaghetti　butter　tomato

tomato sauce

onion

rosemary

chopped garlic

bay-leaf

❷ Match the words and phrases with their meaning.

flour	迷迭香
white pepper	意大利面
spaghetti	番茄酱汁
tomato sauce	香叶
bay-leaf	面粉
rosemary	白胡椒

❸ What's this?

It's____.

It is made of____________________________________.

Words List

面粉	flour ['flaʊə] n.
意大利面条	spaghetti [spə'getɪ] n.

VI Coppiette 罗马式炸小肉饼

❶ Learn the following words and phrases.

minced garlic　tomato paste　rosemary powder

fat pork　beef　onion

nutmeg　pepper　brown sauce

oil　salt　parsley

raisin　milk　toast

flour　egg　bread crumb

spaghetti　pine nut　cheese powder

❷ Match the words and phrases with their meaning.

beef	豆蔻粉
fat pork	蒜蓉
minced garlic	迷迭香粉
tomato paste	牛肉
rosemary powder	番茄膏
nutmeg	肥猪肉
raisins	意大利面
toast	面包屑
flour	葡萄干
bread crumb	松子
spaghetti	鸡蛋
pine nut	烤面包
egg	面粉

❸ What's this?

It's________________________________.

It is made of___________________________.

❹ Classify the following words or phrases.

beef	chopped garlic	tomato paste	rosemary powder
nutmeg	fat pork	toast	bread crumb
brown sauce	pepper	parsley	onion

Sauce（调味汁）

____

____

____

Spices（香料）

____

____

____

Vegetable（蔬菜）

____

____

____

Meat（肉类）

____

____

____

Bread（面包）

____

____

____

Words List

肥的；胖的	fat [fæt] adj.
肥猪肉	fat pork [fæt pɔ:k]
面团；糨糊	paste [peɪst] n.
番茄膏	tomato paste [tə'mɑ:təʊ peɪst]
迷迭香粉	rosemary powder ['rəʊzm(ə)rɪ 'paʊdə]
[植] 肉豆蔻（树）	nutmeg ['nʌtmeg] n.
葡萄干	raisin ['reɪz(ə)n] n.
烤面包	toast [təʊst] n.
碎屑（尤指面包屑或糕饼屑）	crumb [krʌm] n.
面包屑	bread crumb [bred krʌm]
意大利面条	spaghetti [spə'getɪ] n.
松树	pine [paɪn] n.
坚果，坚果果仁	nut [nʌt] n.
松果	pine nut [paɪn nʌt]
奶酪	cheese [tʃi:z] n.

If I had a thousand sons, the first human principle I would teach them would be to forswear thin potations and dedicate themselves to Sherry.

William Shakespeare

词汇在线

Unit 14

Lamb Cooking 羊肉的烹制

在线选择

I Vanilla Baked Lamp Chop and Red Wine Sauce 香草烤羊排配红酒汁

❶ Learn the following words and phrases.

lamb chop　onion　ginger

garlic　rosemary cut　honey

lea perrins　lemon juice　red wine

❷ Match the words and phrases with their meaning.

lamb chop	蜂蜜
ginger	柠檬汁
lemon juice	姜
lea perrins	羊排
honey	迷迭香碎
rosemary cut	李派林

❸ What's this?

It's____.

It is made of___________________________________.

Words List

羔羊，小羊	lamb [læm] n.
姜	ginger ['dʒɪndʒə] n.
切口；切片	cut [kʌt] n.
迷迭香碎	rosemary cut ['rəʊzm(ə)rɪ kʌt]

II Grilled Mutton with Nut and Garlic 果仁香蒜羊排

❶ Learn the following words and phrases.

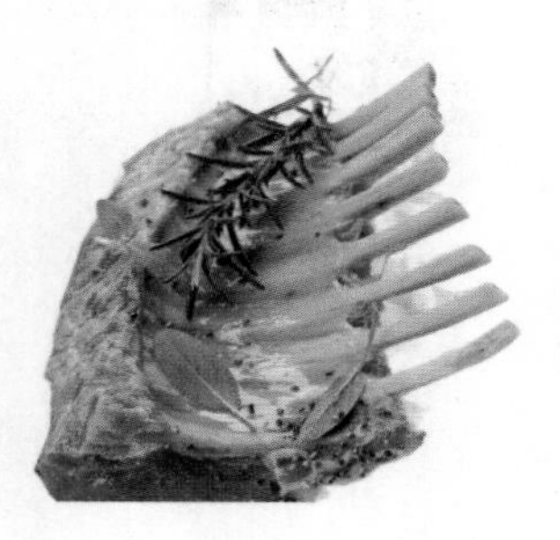

lamp chop

cashew

flour

egg white pepper salt

lea perrins garlic powder red wine

❷ Match the words and phrases with their meaning.

lamb chop	腰果
cashew	蒜香粉
garlic powder	羊排

❸ What's this?

It's____.

It is made of____________________________________.

Words List

羊肉；绵羊	mutton ['mʌt(ə)n] n.
羊排	mutton chop ['mʌt(ə)n tʃɒp]
腰果，腰果树	cashew ['kæʃu:; kə'ʃu:] n.
蒜香粉	garlic powder ['gɑ:lɪk 'paʊdə]

中文菜名英文译法 ③

以烹制方法为主，原料为辅

类别	中文菜名示例	英文译法示例	翻译原则
做法＋主料	拌双耳	Tossed Black and White Fungus	做法(动词过去分词) + 主料(名称/形状)
做法＋主料＋配料	豌豆辣牛肉	Sautéed Spicy Beef and Green Peas	做法(动词过去分词) + 主料(名称/形状)+ 配料
做法＋主料＋汤汁	川北凉粉	Tossed Clear Noodles with Chili Sauce	做法(动词过去分词) + 主料(名称/形状)+ with /in + 汤汁

以形状、口感为主，原料为辅

类别	中文菜名示例	英文译法示例	翻译原则
形状/口感＋主配料	玉兔馒头	Rabbit-Shaped Mantou	形状+主料
	脆皮鸡	Crispy Chicken	口感+主料
做法＋形状/口感＋主料＋配料	小炒黑山羊	Sautéed Sliced Lamb with Pepper and Parsley	做法(动词过去分词) + 形状/口感+ with + 主料 + 配料

介词in和with在汤汁、配料中的用法

类别	中文菜名示例	英文译法示例	翻译原则
主料是浸在汤汁或配料中	豉汁牛仔骨	Steamed Beef Ribs in Black Bean Sauce	主料是浸在汤汁或配料中时，用in连接
汤汁或蘸料与主料是分开的，或是后浇在主菜上的	泡椒鸭丝	Shredded Duck with Pickled Peppers	汤汁或蘸料和主料是分开的，或是后浇在主菜上的，则用with连接

词汇在线

Unit 15

Fish and Shellfish Cooking 鱼类和贝类的烹制

在线选择

I Grilled Prawn 扒大虾

❶ Learn the following words and phrases.

❷ Match the words and phrases with their meaning.

prawn	莳萝
mashed potato	菊苣
purple leaf lettuce	芝麻菜
endive	土豆泥
arugula	明虾
dill	紫叶生菜

❸ What's this?

It's____.

It is made of__________________________________.

Words List

明虾	prawn [prɔ:n] n.
莳萝；草茴香	dill [dɪl] n.
紫叶生菜	purple leaf lettuce ['pɜ:p(ə)l li:f 'letɪs]
菊苣	endive ['endaɪv; -dɪv] n.
芝麻菜	arugula [æ'ru:gjʊlə] n.

II Steamed Sole Fillet with Mashed Potatoes, Tomato and Basil Sauce 清蒸龙利鱼柳配土豆泥、番茄汁和紫苏汁

❶ Learn the following words and phrases.

❷ Match the words and phrases with their meaning.

lemon	橄榄油
sole	柠檬
sweet bean	龙利鱼
olive oil	紫苏
basil	甜豆

❸ What's this?

It's____.

It is made of____________________________________.

Words List

龙利鱼	sole [səʊl] n.
荷兰豆；甜豆	sweet bean [swi:t bi:n]
紫苏，罗勒	basil ['bæzl] n.

III Boiled Fish Fillet with White Wine 白葡萄酒汁煮鱼柳

❶ Learn the following words and phrases.

celery　onion　carrot

silvery pout　bay-leaf　lemon

❷ Match the words and phrases with their meaning.

bay-leaf	蘑菇
sole	香叶
mushroom	龙利鱼
silvery pout	菠菜
spinach	银鳕鱼

❸ What's this?

It's____________________________________.

It is made of________________________________.

Words List

银鳕鱼 silvery pout ['sɪlv(ə)rɪ paʊt] n.
蘑菇 mushroom ['mʌʃruːm; -rʊm] n.
菠菜 spinach ['spɪnɪdʒ; -ɪtʃ] n.

IV Fried Seafood 炸海鲜

❶ Learn the following words and phrases.

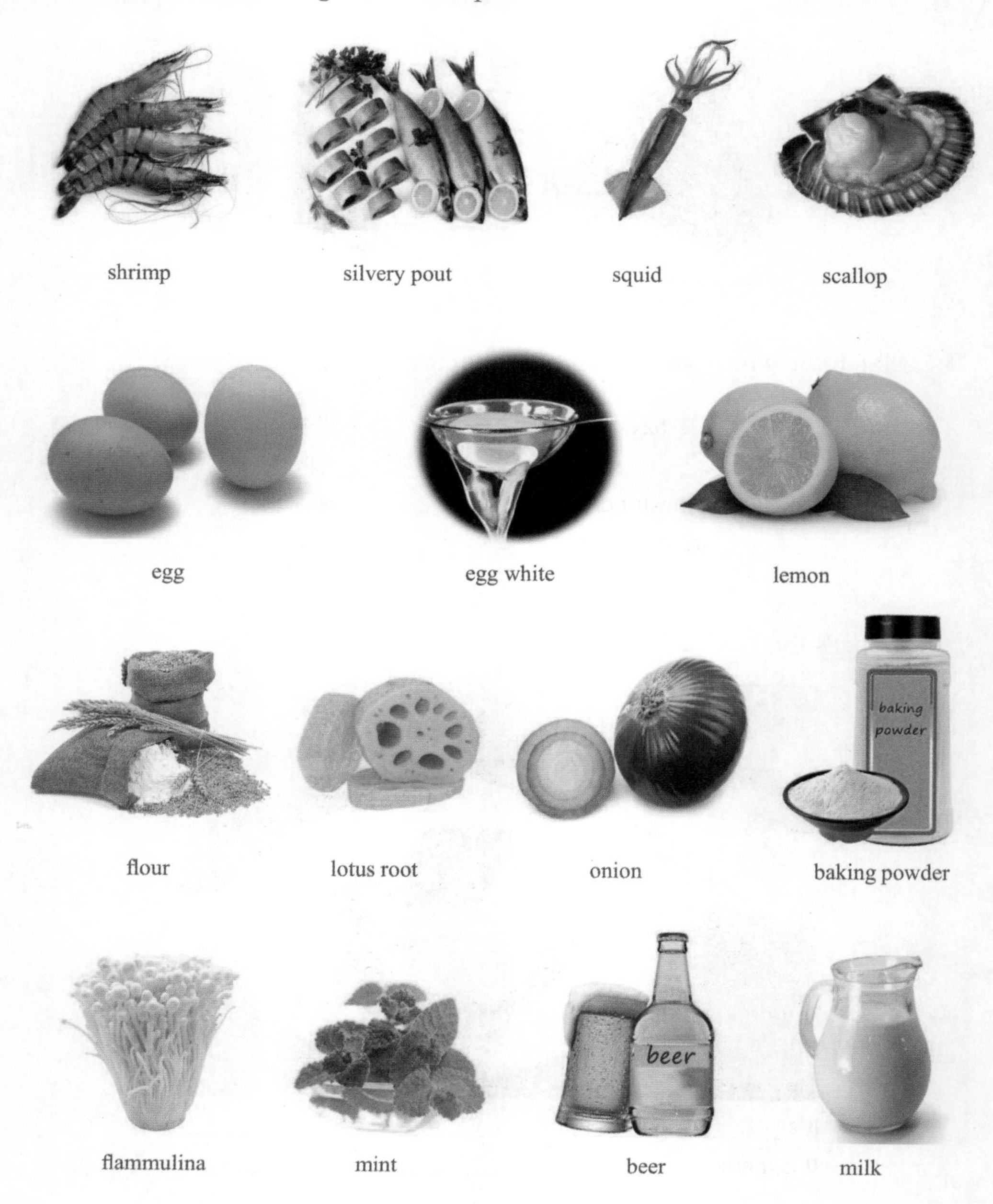

❷ Match the words and phrases with their meaning.

squid	扇贝
silvery pout	薄荷
lotus root	鱿鱼
scallop	鸡蛋清
mint	银鳕鱼
egg white	泡打粉
baking powder	莲藕

❸ What's this?

It's____.

It is made of______________________________________.

Words List

扇贝	scallop ['skæləp] n.
鱿鱼	squid [skwɪd] n.
金针菇	flammulina ['flæmjʊlɪnə] n.
莲藕	lotus root ['ləʊtəs ru:t]
[食品]发酵粉	baking powder ['beɪkɪŋ 'paʊdə]

V Stuffed Sliced Fish Fillet 卷酿鱼柳

❶ Learn the following words and phrases.

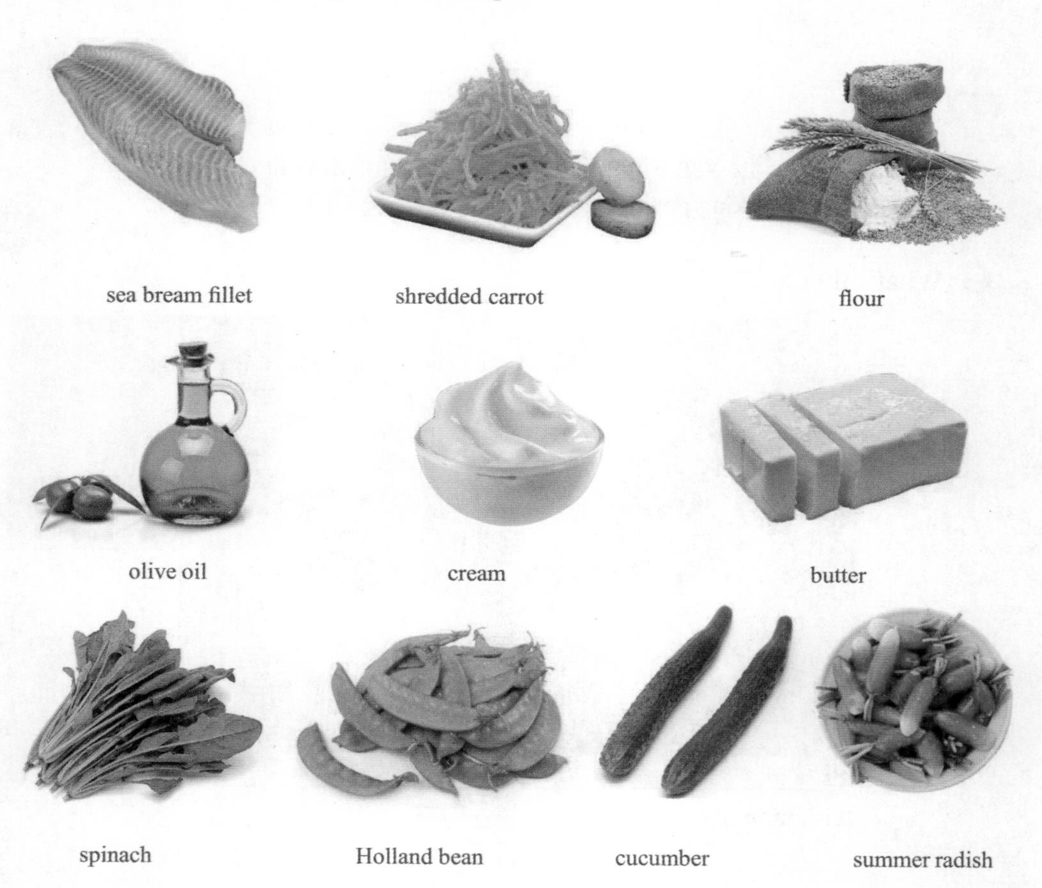

❷ Match the words and phrases with their meaning.

cucumber	胡萝卜丝
spinach	荷兰豆
sea bream fillet	菠菜
Holland bean	黄瓜
flour	黄油
butter	水萝卜
shredded carrot	鲷鱼柳
summer radish	面粉

3 What's this?

It's____.

It is made of__________________________________.

Words List

鲷鱼	sea bream [si: bri:m]
胡萝卜丝	shredded carrot ['ʃredɪd 'kærət]
荷兰豆	Holland bean ['hɒlənd bi:n]
水萝卜	summer radish ['sʌmə 'rædɪʃ]

VI Fried Silver Pout 煎银鳕鱼

1 Learn the following words and phrases.

silvery pout

lemon

olive oil

crab mushroom

sweet bean

cherry radish

purple leaf lettuce　　dill　　basil

endive　　capers

❷ Match the words and phrases with their meaning.

sweet bean	蟹味菇
silvery pout	水瓜柳
cherry radish	银鳕鱼
purple leaf lettuce	甜豆
endive	紫叶生菜
capers	樱桃萝卜
crab mushroom	菊苣

❸ What's this?

It's____.

It is made of____________________________________.

Words List

银鳕鱼	silvery pout ['sɪlv(ə)rɪ paʊt]
蟹味菇	crab mushroom [kræb 'mʌʃruːm]
樱桃萝卜	cherry radish ['tʃerɪ 'rædɪʃ]
水瓜柳；酸豆	capers ['keɪps] n.

VII Grilled Salmon 扒三文鱼排

❶ Learn the following words and phrases.

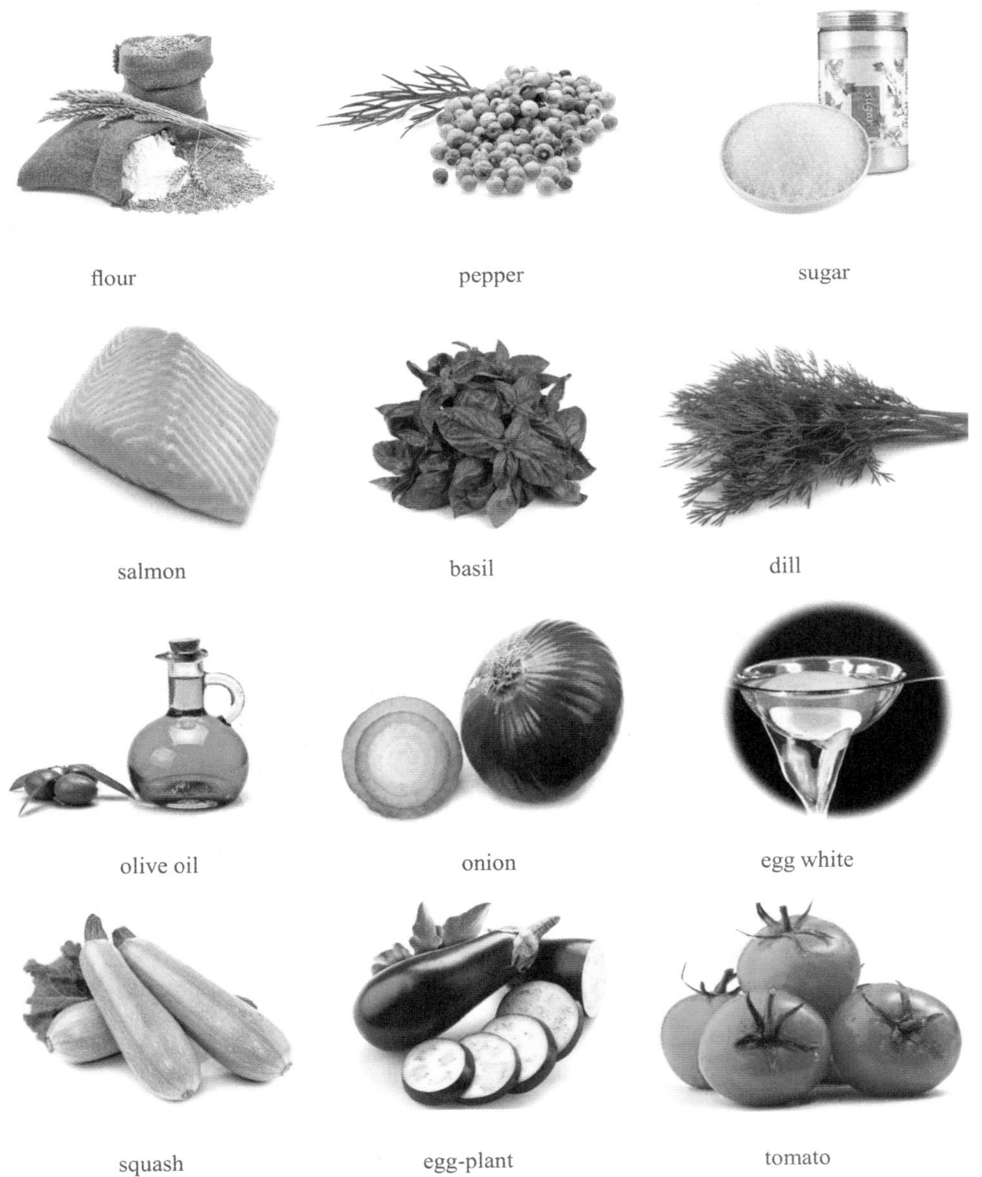

flour pepper sugar

salmon basil dill

olive oil onion egg white

squash egg-plant tomato

ketchup

thyme

balsamico

❷ Match the words and phrases with their meaning.

squash	三文鱼
dill	百里香
salmon	西葫芦
egg white	意大利黑醋
egg-plant	莳萝
balsamico	罗勒
basil	茄子
thyme	蛋清

❸ What's this?

It's____.

It is made of__________________________________.

Words List

三文鱼；鲑鱼	salmon ['sæmən] n.
莳萝	dill [dɪl] n.
西葫芦	squash [skwɒʃ] n.
茄子	egg-plant [eg plɑ:nt] n.
百里香	thyme [taɪm] n.
意大利黑醋	balsamico [bɒl'sæmɪkə] n.

中文菜名英文译法 ❹

以人名、地名为主，原料为辅

类别	中文菜名示例	英文译法示例	翻译原则
创始人(发源地)＋主料	麻婆豆腐	Mapo Doufu (Sautéed Doufu in Hot and Spicy Sauce)	人名 + 主料
	广东点心	Cantonese Dim Sum	地名 + 主料
创始人(发源地)＋主配料＋做法	四川辣子鸡	Spicy Chicken, Sichuan Style	做法(动词过去式) + 主辅料 +人名/地名 + Style
	北京炸酱面	Noodles with Soy Bean Paste, Beijing Style	

词汇在线

Unit 16

Poultry Cooking 禽类的烹制

在线选择

I Roasted Spring Chicken with Vanilla 香草烤雏鸡

❶ Learn the following words and phrases.

spring chicken rosemary thyme

fennel seed sauce Brandy

tomato garlic

❷ Match the words and phrases with their meaning.

vanilla	小茴香粉
spring chicken	百里香
rosemary	浇汁
thyme	迷迭香
sauce	香草
fennel seed	童子鸡

❸ What's this?

It's____.

It is made of____________________________________.

Words List

香子兰，香草	vanilla [və'nɪlə] n.
春季；泉水	spring [sprɪŋ] n.
童子鸡	spring chicken [sprɪŋ 'tʃɪkɪn]
[植] 茴香；小茴香	fennel ['fen(ə)l] n.
种子，子孙	seed [si:d] n.
小茴香子	fennel seed ['fen(ə)l si:d]

II Stewed Duck with Orange Juice 橙汁烩鸭

❶ Learn the following words and phrases.

duck breast meat

orange

brown sauce

asparagus

mushroom

celery

carrot

onion

❷ Match the words and phrases with their meaning.

duck breast meat	芦笋
orange	芹菜
asparagus	橙子
celery	鸭脯肉
carrot	胡萝卜

❸ What's this?

It's______________________________________.

It is made of__________________________________.

Words List

鸭子，鸭肉	duck [dʌk] n.
乳房，乳腺；胸脯，胸部	breast [brest] n.
鸭脯肉	duck breast meat [dʌk brest mi:t]
橘子，橙子	orange ['ɒrɪn(d)ʒ] n.

III Christmas Turkey 圣诞烤火鸡

❶ Learn the following words and phrases.

turkey onion celery

carrot mushroom sauce cranberry sauce

❷ Match the words and phrases with their meaning.

Christmas	小红莓果酱
turkey	蘑菇汁
mushroom sauce	圣诞节
cranberry sauce	火鸡

❸ What's this?

It's____.

It is made of______________________________________.

Words List

圣诞节	Christmas ['krɪsməs] n.
火鸡	turkey ['tɜːkɪ] n.
蘑菇汁	mushroom sauce ['mʌʃruːm sɔːs]
[植] 蔓越橘；小红梅	cranberry ['krænb(ə)rɪ] n.
小红梅果酱	cranberry sauce ['krænb(ə)rɪ sɔːs]

IV Stewed Quail with Red Wine 红酒烩鹌鹑

❶ Learn the following words and phrases.

❷ Match the words and phrases with their meaning.

quail	橄榄油
bay-leaf	芹菜
white pepper	鹌鹑
carrot	香叶
celery	白胡椒粉
olive oil	胡萝卜

❸ What's this?

It's____.

It is made of____________________________________.

Words List

鹌鹑；鹌鹑肉 quail [kweɪl] n.

V Curried Chicken 咖喱鸡

❶ Learn the following words and phrases.

coconut milk lemongrass galangal

chilli ginger root chopped garlic

❷ Match the words and phrases with their meaning.

coconut milk	香蕉
lemongrass	干辣椒
galangal	大蒜米
chilli	苹果
ginger root	椰浆
chopped garlic	香茅草
apple	老姜
banana	生姜

❸ What's this?

It's____.

It is made of__________________________________.

❹ Classify the following words or phrases.

coconut　lemongrass　galangal　chilli
ginger root　apple　chopped garlic
mushroom　banana　potato　carrot

Fruit（水果）	Condiment（调味料）	Vegetable（蔬菜）
________	________	________
________	________	________
________	________	________

Words List

香茅；柠檬草	lemongrass ['l məngræs] n.
根，根源	root [ru:t] n.
南姜，高良姜	galangal [gə'læŋgəl] n.
老姜	ginger root ['dʒɪndʒə ru:t]

葡萄酒的颜色鉴别

白葡萄酒

青绿　淡黄　铜绿　淡金　黄铜

红葡萄酒

宝石红　紫红　紫色　深红　砖红

Module
模块 4

词汇在线

在线选择

Unit 17

Pasta 意大利面食

I Spicy Chicken Pizza 香辣烤鸡比萨

❶ Learn the following words and phrases.

Ⓐ Pizza Dough

high gluten flour

olive oil

yeast

salt

water

Ⓑ Ketchup

tomato

tomato paste

ketchup

bay-leaf

hyssop

pepper

sugar

Ⓒ Spicy Chicken Pizza

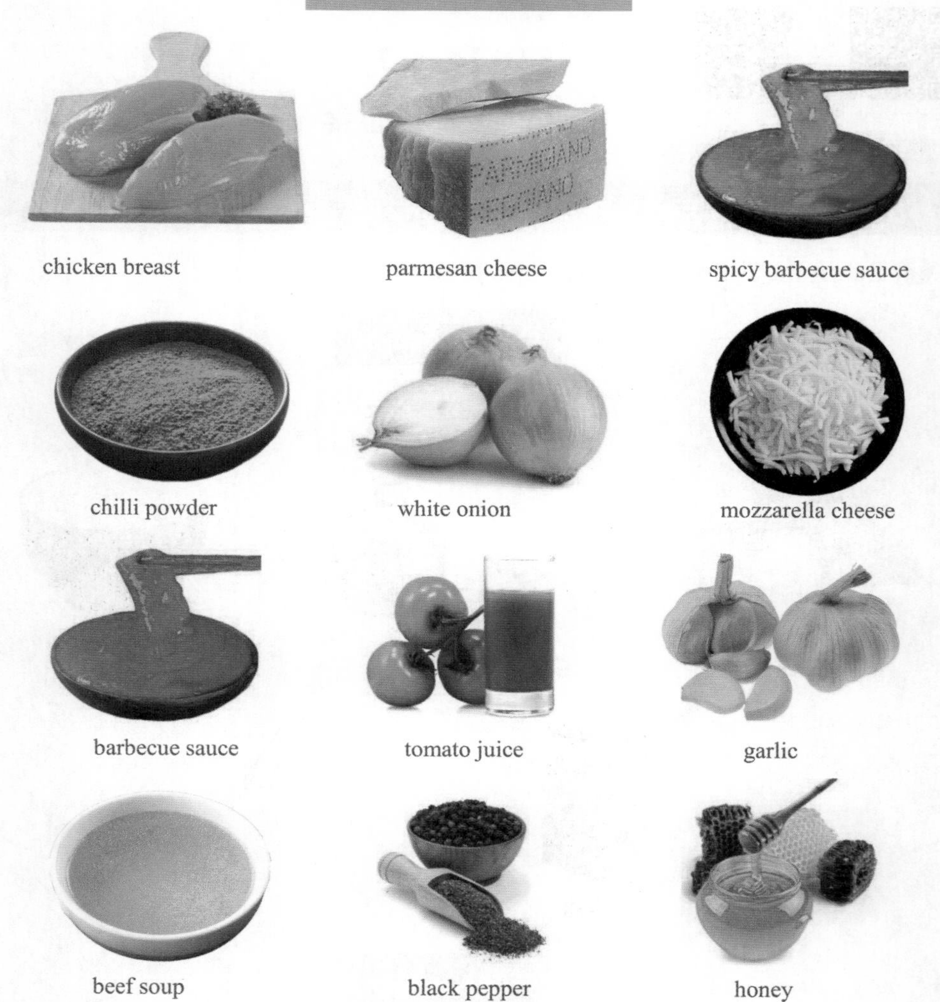

chicken breast　parmesan cheese　spicy barbecue sauce

chilli powder　white onion　mozzarella cheese

barbecue sauce　tomato juice　garlic

beef soup　black pepper　honey

❷ Match the words and phrases with their meaning.

high gluten flour	西红柿
olive oil	番茄酱
yeast	番茄膏
tomato	橄榄油
tomato paste	香叶
ketchup	高筋粉
bay-leaf	胡椒
hyssop	牛膝草
pepper	酵母

B

chicken breast	白洋葱
parmesan cheese	马苏里拉奶酪
spicy barbecue sauce	大蒜
chilli powder	番茄汁
white onion	鸡胸肉
mozzarella cheese	帕玛森奶酪
barbecue sauce	牛肉浓汤
tomato juice	蜂蜜
garlic	黑胡椒碎
beef soup	香辣烧烤酱
black pepper	辣椒粉
honey	烧烤酱

❸ What's this?

It's______________________________________.

It is made of__________________________________.

Words List

比萨	pizza ['pi:tsə] n.
比萨面团	pizza dough ['pi:tsə dəʊ]
高筋粉	high gluten flour [haɪ 'glu:t(ə)n 'flauə]
酵母	yeast [ji:st] n.
牛膝草	hyssop ['hɪsəp]
奶酪；干酪	cheese [tʃi:z] n.
帕玛森奶酪	parmesan cheese [ˌpa:mi'zæn tʃi:z]
辣的	spicy ['spaɪsɪ] adj.

烧烤(B.B.Q)　barbecue ['bɑ:bɪkju:] vt. n.
烧烤酱　barbecue sauce ['bɑ:bɪkju: sɔ:s]
马苏里拉奶酪　mozzarella cheese ['mɒtsə'relə tʃi:z]
番茄酱　ketchup ['ketʃəp] n.

II Ham and Pineapple Pizza 火腿凤梨比萨

❶ Learn the following words and phrases.

mozzarella cheese　parmesan cheese　ham

pineapple　marjoram

❷ Match the words and phrases with their meaning.

mozzarella cheese	火腿
parmesan cheese	菠萝
ham	马苏里拉奶酪
pineapple	帕玛森奶酪
marjoram	甘牛至

❸ What's this?

It's______________________________________.

It is made of__________________________________.

Words List

菠萝；凤梨　pineapple ['paɪnæp(ə)l] n.

甘牛至；马佐林　marjoram ['mɑːdʒ(ə)rəm] n.

III Seafood Pizza 海鲜比萨

❶ Learn the following words and phrases.

mussel　shrimp　scallop

parsley　mozzarella cheese　chilli powder

octopus　capers　water　white wine

❷ Match the words and phrases with their meaning.

mozzarella cheese	水瓜柳
shrimp	欧芹
scallop	八爪鱼
parsley	马苏里拉奶酪
mussel	虾
chilli powder	白酒
octopus	水
capers	辣椒粉
water	带子
white wine	青口贝

❸ What's this?

It's____.

It is made of__________________________________.

Words List

带子；扇贝	scallop ['skæləp] n.
蚌；青口贝	mussel ['mʌsl] n.
章鱼；八爪鱼	octopus ['ɔktəpəs] n.
水瓜柳；酸豆	capers ['keipəz] n.

IV Spaghetti Bolognaise 茄汁肉酱意大利面

❶ Learn the following words and phrases.

high gluten noodles　bolognaise　butter

tomato juice　hyssop　pepper　salt

cheese powder

oil

chicken powder

❷ Match the words and phrases with their meaning.

high gluten noodles	胡椒
bolognaise	鸡粉
butter	油
tomato juice	盐
oil	黄油
pepper	高筋面条
cheese powder	番茄汁
salt	牛膝草
chicken powder	奶酪粉
hyssop	番茄肉酱

❸ What's this?

It's____.

It is made of____________________________________.

Words List

意大利面食	pasta ['pæstə] n.
番茄肉酱	bolognaise [ˌbəʊlɔg'naɪs] n.
高筋面条	high gluten noodles [haɪ 'glu:t(ə)n 'nu:dlz]
奶酪粉	cheese powder [tʃi:z 'paʊdə]

V Lasagne Verde Al Forno 焗意大利青面

❶ Learn the following words and phrases.

beef paste　high gluten flour　spinach

egg　salt　red wine

pepper　tomato sauce　emmental

chicken powder　cream sauce

❷ Match the words and phrases with their meaning.

beef paste	鸡蛋
high gluten flour	红酒
spinach	盐
egg	番茄酱汁
salt	奶油汁
red wine	瑞士干酪
pepper	高筋面粉
tomato sauce	鸡粉
emmental	菠菜
chicken powder	胡椒
cream sauce	牛肉酱

❸ What's this?

It's____.

It is made of__________________________________.

Words List

焗意大利青面 lasagne verde al forno [lə'zænjə vɜːd æl 'fɔːnəʊ]

牛肉酱 beef paste [biːf peɪst]

瑞士（多孔）干酪；埃曼 塔奶酪 emmental ['eməntaːl] n.

奶油汁 cream sauce [kriːm sɔːs]

词汇在线

在线选择

Unit 18

Bread 面包

I Soft Bread 软面包

❶ Learn the following words and phrases.

bread flour　condensed milk　egg pulp

sugar　butter　ameliorant

salt　water　yeast

❷ Match the words and phrases with their meaning.

condensed milk	面包粉
sugar	酵母
yeast	清水
water	盐
bread flour	改良剂
butter	鸡蛋液
salt	黄油
egg pulp	糖
ameliorant	炼乳

❸ What's this?

It's ____.

It is made of ____________________________________.

Words List

软的	soft [sɔft] adj.
面包粉	bread flour [bred 'flauə]
炼乳	condensed milk [kən'denst milk]
鸡蛋液	egg pulp [eg pʌlp]
改良剂	ameliorant [ə'mi:ljərənt] n.

II Bread Stick 面包棍

❶ Learn the following words and phrases.

bread flour　　yeast　　ameliorant

water　　salt

❷ Match the words and phrases with their meaning.

salt	面包粉
ameliorant	酵母
yeast	清水
bread flour	盐
water	改良剂

❸ What's this?

It's____________________________________

______________________________________.

It is made of______________________________

______________________________________.

Words List

面包粉	bread flour [bred 'flauə]
酵母	yeast [ji:st] n.
改良剂	ameliorant [ə'mi:ljərənt] n.
棍；手杖	stick [stɪk] n.

III Croissant 牛角包

❶ Learn the following words and phrases.

bread flour

yeast

salt

milk

butter

sugar

shortening

❷ Match the words and phrases with their meaning.

salt	牛奶
yeast	糖
bread flour	盐
milk	黄油
shortening	起酥油
sugar	酵母
butter	面包粉

❸ What's this?

It's____.

It is made of__________________________________.

Words List

牛角包 croissant [krwaːsɒŋ] n.

牛奶 milk [mɪlk] n.

起酥油 shortening ['ʃɔːt(ə)nɪŋ] n.

IV Toast 烤面包片

❶ Learn the following words and phrases.

❷ Match the words and phrases with their meaning.

bread flour	奶粉
yeast	糖
milk powder	起酥油
sugar	盐
water	鸡蛋
salt	水
egg	酵母
ameliorant	面包粉
shortening	改良剂
baking powder	[食品]发酵粉

③ What's this?

It's______________________________________.

It is made of________________________________.

Words List

烤面包片	toast [təʊst] n.
奶粉	milk powder [mɪlk 'paʊdə]
[食品]发酵粉	baking powder ['beɪkɪŋ 'paʊdə]

V Doughnut 甜甜圈

① Learn the following words and phrases.

bread flour　salt　condensed milk　water

sugar　butter　yeast　egg pulp

❷ Match the words and phrases with their meaning.

condensed milk	面包粉
sugar	酵母
yeast	清水
water	盐
butter	鸡蛋液
salt	黄油
egg pulp	糖
bread flour	炼乳

❸ What's this?

It's____________________________________.

It is made of______________________________.

Words List

甜甜圈 doughnut ['dəunʌt] n.

词汇在线

在线选择

Unit 19

Cake 蛋糕

I Sponge Cake 海绵蛋糕

❶ Learn the following words and phrases.

cake mix　　cake emulsifier　　water

sugar　　salad oil　　egg pulp

❷ Match the words and phrases with their meaning.

salad oil	清水
egg pulp	蛋糕粉
sugar	糖
cake mix	鸡蛋液
water	蛋糕油
cake emulsifier	色拉油

❸ What's this?

It's____.

It is made of______________________________.

Words List

海绵蛋糕	sponge cake [spʌndʒ keɪk]
蛋糕油，蛋糕乳化剂	cake emulsifier [keɪk ɪ'mʌlsɪfaɪə]
蛋糕粉	cake mix [keɪk mɪks]
色拉油	salad oil ['sæləd ɒɪl]

II Chiffon Cake 戚风蛋糕

❶ Learn the following words and phrases.

cake mix

baking powder

cream of tartar

powdered sugar

milk

egg yolk

egg white

salad oil

salt

vanilla extract

❷ Match the words and phrases with their meaning.

egg yolk	蛋糕粉
powdered sugar	色拉油
cream of tartar	蛋白
cake mix	糖粉
baking powder	牛奶
salt	蛋黄
salad oil	发酵粉
vanilla extract	香草精
milk	塔塔粉
egg white	盐

❸ What's this?

It's____.

It is made of__________________________________.

Words List

戚风蛋糕	chiffon cake ['ʃɪfɒn keɪk]
蛋黄液	egg yolk [eg jəʊk]
香草精	vanilla extract [və'nɪlə 'ekstrækt]
糖粉	powdered sugar ['paʊdəd 'ʃʊgə]
塔塔粉	cream of tartar [kri:m ɒv 'tɑ:tə]
蛋清液	egg white [eg waɪt]
[食品]发酵粉	baking powder ['beɪkɪŋ 'paʊdə]

III Pound Cake 重油蛋糕

❶ Learn the following words and phrases.

cake mix　baking powder　salt　powdered sugar

milk　egg pulp　butter　vanilla extract

❷ Match the words and phrases with their meaning.

egg pulp	蛋糕粉
powdered sugar	黄油
cake mix	糖粉
baking powder	牛奶
salt	全蛋液
butter	[食品] 发酵粉
vanilla extract	香草精
milk	盐

③ What's this?

It's____.

It is made of___________________________________.

Words List

重油蛋糕　pound cake [paʊnd keɪk]

IV Fruit Cake 水果蛋糕

① Learn the following words and phrases.

salt　egg pulp　butter　vanilla extract

fruit　all purpose flour

② Match the words and phrases with their meaning.

egg pulp	香草精
salt	鸡蛋液
butter	水果
vanilla extract	盐
fruit	中筋粉
all purpose flour	黄油

③ What's this?

It's____.

It is made of___________________________________.

Words List

水果蛋糕 fruit cake [fru:t keɪk]

中筋粉 all purpose flour [ɔ:l 'pɜ:pəs 'flaʊə]

V Mousse 慕斯

① Learn the following words and phrases.

chiffon cake billet　　berry sugar　　whipping cream　　gelatine

fruit puree of strawberry

chocolate

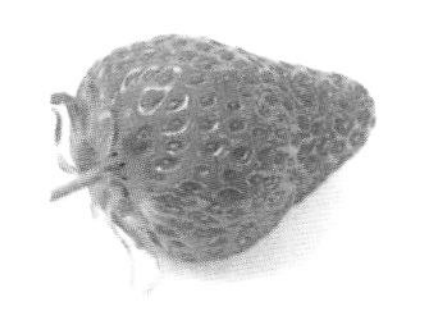

strawberry

cream

❷ Match the words and phrases with their meaning.

chiffon cake billet	淡奶油
berry sugar	巧克力
whipping cream	草莓
gelatine	戚风蛋糕坯
fruit puree of strawberry	鲜奶油
chocolate	明胶，胶质
strawberry	草莓果蓉
cream	细砂糖

❸ What's this?

It's____.

It is made of____________________________________.

Words List

慕斯	mousse [muːs] n.
戚风蛋糕坯	chiffon cake billet ['bɪlɪt]
巧克力	chocolate ['tʃɒk(ə)lət]
草莓	strawberry ['strɔːb(ə)rɪ] n.
鲜奶油	cream [kriːm] n.
胶质；果子冻；白明胶	gelatine ['dʒ lətn] n.
草莓果蓉	fruit puree of strawberry [fruːt 'puːrɪ ɒv 'strɔːb(ə)rɪ]

VI Tiramisu Cake 提拉米苏

❶ Learn the following words and phrases.

❷ Match the words and phrases with their meaning.

wafer biscuit	乳酪
ground coffee	蛋黄
boiling water	鲜奶油
cointreau	可可粉
sugar	威化饼
cheese	清水
egg yolk	热开水
cocoa powder	君度酒
whipping cream	糖
water	咖啡粉
gelatine	明胶

❸ What's this?

It's____.

It is made of____.

Words List

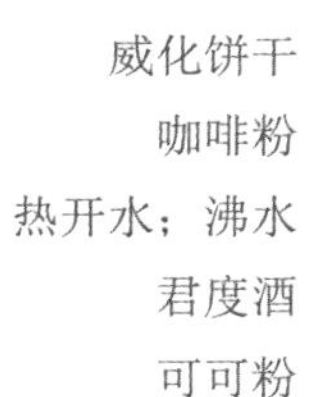

提拉米苏	tiramisu cake [ˌtɪrəmɪ'suː keɪk]
威化饼干	wafer biscuit ['weɪfə 'bɪskɪt]
咖啡粉	ground coffee [graʊnd 'kɒfɪ]
热开水；沸水	boiling water ['bɒɪlɪŋ 'wɔːtə]
君度酒	cointreau [kwɒn'tro] n.
可可粉	cocoa powder ['kəʊkəʊ 'paʊdə]

词汇在线

在线选择

Unit 20

Other 其他甜点

I Cookies 小甜饼/曲奇

❶ Learn the following words and phrases.

whipping cream　berry sugar　salt

vanilla extract　all purpose flour　egg pulp

❷ Match the words and phrases with their meaning.

whipping cream	鸡蛋液
berry sugar	中筋粉
salt	细砂糖
egg pulp	香草精
all purpose flour	盐
vanilla extract	鲜奶油

③ What's this?

It's______________________________________.

It is made of______________________________________.

Words List

饼干 cookie ['kʊkɪ] n.

鲜奶油 whipping cream ['wɪpɪŋ kriːm]

细砂糖 berry sugar ['berɪ 'ʃʊgə]

II Pancake 薄饼

① Learn the following words and phrases.

butter

low gluten flour

shredded coconut stuffing

powdered sugar

egg white

❷ Match the words and phrases with their meaning.

butter	低筋粉
low gluten flour	椰蓉
egg white	糖粉
powdered sugar	黄油
shredded coconut stuffing	蛋白

❸ What's this?

It's____.

It is made of____________________________________.

Words List

薄饼 pancake ['pænkeɪk] n.

低筋粉 low gluten flour [ləʊ 'glu:t(ə)n 'flaʊə]

椰蓉 shredded coconut stuffing [ʃredid 'kəʊkənʌt 'stʌfɪŋ]

III Egg Tart 蛋塔

❶ Learn the following words and phrases.

flour

butter

water

❷ Match the words and phrases with their meaning.

flour	鸡蛋
butter	面粉
water	开水
salt	清水
egg	黄油
boiled water	玉米淀粉
sugar	蛋奶香精
milk	盐
corn starch	糖
egg-milk flavour	牛奶

❸ What's this?

It's____.

It is made of___________________________________.

Words List

蛋塔 egg tart [eg tɑ:t]

玉米淀粉 corn starch [kɔ:n stɑ:tʃ]

蛋奶香精 egg-milk flavour [eg mɪlk 'fleɪvə]

IV Muffin 松饼

❶ Learn the following words and phrases.

baking powder　　egg　　honey

milk　　oil　　pancake mix

powdered sugar　　strawberry

❷ Match the words and phrases with their meaning.

baking powder	蜂蜜
egg	糖粉
honey	草莓
milk	松饼粉
oil	油
powdered sugar	鸡蛋
strawberry	牛奶
pancake mix	［食品］发酵粉

③ What's this?

It's____.

It is made of____________________________________.

Words List

小松饼	muffin ['mʌfɪn] n.
蜂蜜	honey ['hʌnɪ] n.
西式煎饼粉	pancake mix ['pænkeɪk mɪks]

V Puff 泡芙

❶ Learn the following words and phrases.

flour   cream   egg pulp

water   powdered sugar

❷ Match the words and phrases with their meaning.

flour	水
cream	鸡蛋液
egg pulp	面粉
water	糖粉
powdered sugar	奶油

❸ What's this?

It's____.

It is made of____________________________________.

Words List

泡芙 puff [pʌf] n.

VI Pudding 布丁

❶ Learn the following words and phrases.

low gluten flour　　sugar　　baking powder

butter　　　egg pulp　　　milk

❷ Match the words and phrases with their meaning.

low gluten flour	黄油
sugar	［食品］发酵粉
baking powder	鸡蛋液
butter	低筋粉
egg pulp	牛奶
milk	糖

❸ What's this?

It's____.

It is made of__________________________________.

Words List

布丁　pudding [ˈpʊdɪŋ] n.

VII Ice Cream 冰激凌

❶ Learn the following words and phrases.

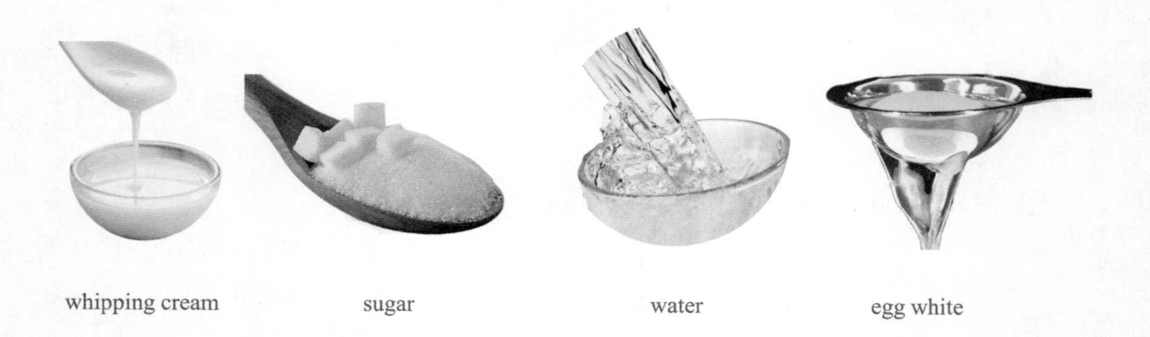

❷ Match the words and phrases with their meaning.

whipping cream	水
sugar	蛋清液
water	糖
egg white	鲜奶油

❸ What's this?

It's____.

It is made of__________________________________.

词汇在线

Unit 21

Chocolate and Gingerbread 巧克力和姜饼

在线选择

I Chocolate 巧克力

❶ Learn the following words and phrases.

❷ Match the words and phrases with their meaning.

milk	可可粉
cocoa powder	果仁
cocoa butter	可可脂
nutlet	细砂糖
berry sugar	牛奶

❸ What's this?

It's____.

It is made of__________________________________.

Words List

巧克力	chocolate ['tʃɒk(ə)lət] n.
可可脂	cocoa butter ['kəʊkəʊ 'bʌtə]
果仁	nutlet ['nʌtlɪt] n.

II Gingerbread 姜饼

❶ Learn the following words and phrases.

shortening　powdered sugar　egg pulp　low gluten flour

baking soda　ground turmeric　clove　molasses

cinnamon

salt

❷ Match the words and phrases with their meaning.

shortening	低筋粉
powdered sugar	苏打粉
egg pulp	盐
low gluten flour	起酥油
baking soda	鸡蛋液
ground turmeric	肉桂粉
molasses	丁香碎
cinnamon	糖粉
salt	糖蜜
clove	姜黄粉

❸ What's this?

It's____.

It is made of___________________________________.

Words List

姜饼 gingerbread ['dʒɪndʒəbred] n.

苏打粉 baking soda ['beɪkɪŋ 'səʊdə]

姜黄粉 ground turmeric [graʊnd 'tɜːmərɪk]

糖蜜 molasses [mə'læsɪz] n.

肉桂 cinnamon ['sɪnəmən] n.

丁香碎 clove [kləʊv] n.

中文菜名英文译法 ⑤

体现中国餐饮文化的
用汉语拼音命名或用音译名

中文菜名示例	英文译法示例	翻译原则
饺子	Jiaozi	具有中国特色且被外国人接受的传统食品，本着推广汉语及中国餐饮文化的原则，使用汉语拼音
包子	Baozi	
馒头	Mantou	
花卷	Huajuan	
烧卖	Shaomai	
宫保鸡丁	Kung Pao Chicken	具有中国特色且已被国外主要英文字典收录的，使用汉语方言拼写或音译拼写的菜名，仍保留其原拼写方式
馄饨	Wonton	
佛跳墙	Fotiaoqiang (Steamed Abalone with Shark's Fin and Fish Maw in Broth)	中文菜肴名称无法体现其做法及主配料的，使用汉语拼音，并在其后标注英文注释
锅贴	Guotie (Pan-Fried Dumplings)	
窝头	Wotou (Steamed Corn/Black Rice Bun)	
蒸饺	Steamed Jiaozi (Steamed Dumplings)	
油条	Youtiao (Deep-Fried Dough Sticks)	
汤圆	Tangyuan (Glutinous Rice Balls)	
元宵	Yuanxiao (Glutinous Rice Balls for Lantern Festival)	
粽子	Zongzi (Glutinous Rice Wrapped in Bamboo Leaves)	
驴打滚儿	Lűdagunr (Glutinous Rice Rolls Stuffed with Red Bean Paste)	
豆汁儿	Douzhir (Fermented Bean Drink)	
艾窝窝	Aiwowo (Steamed Rice Cakes with Sweet Stuffing)	

Module

模块 5

- Hotel Menu 酒店菜单
- Dialogues 对话

餐饮英文
译写规范

Unit 22

Hotel Menu 酒店菜单

I Breakfast Menu 早餐菜单

AMERICAN BREAKFAST

Available From 6 a.m. to 11 a.m.

早餐菜单
参考译文

Choice of Fresh Fruit or Vegetable Juice

Orange Watermelon Pineapple
Apple Carrot Cucumber

Combination of Seasonal Fruits

Choice from the Bakery

Croissant Danish Muffin Brioche
Roll White or Wheat Toast

Served with a Selection of Fruit Jams, Honey and Butter

Cereal Selection

Corn Flakes All Bran
Raisin Bran Rice Krispies

Home Made Granola with Skim or Full Cream Milk

Two Fresh Farm Eggs Any Style
With Bacon, Ham, Pork or Chicken Sausage

Served with Hash Brown, Grilled Tomato and Mushrooms

Choice of Freshly Brewed Coffee, Decaffeinated Coffee or Tea

RMB 96

CONTINENTAL BREAKFAST

Available From 6 a.m. to 11 a.m.

Choice of Fresh Fruit or Vegetable Juice

Orange Watermelon Pineapple
Apple Carrot Cucumber

Combination of Seasonal Fruits

Choice from the Bakery

Croissant Danish Muffin Brioche
Roll White or Wheat Toast

Served with a Selection of Fruit Jams, Honey and Butter

Cereal Selection

Corn Flakes All Bran
Raisin Bran Rice Krispies

Home Made Granola with Skim or Full Cream Milk

Choice of Freshly Brewed Coffee, Decaffeinated Coffee or Tea

RMB 96

II Lunch and Dinner Menu 正餐菜单

正餐菜单
参考译文

Available From 11 a.m. to 11 p.m.

STARTERS AND SALADS

Smoked Norwegian Salmon	RMB 78
Beef Carpaccio	RMB 68
Hearts of Romaine	RMB 59
Seasonal Field Greens	RMB 46

SOUPS

Cream Soup	RMB 35
Seafood Soup	RMB 39

SANDWICHES AND BURGERS

Ham and Cheese Sandwich	RMB 78
The Company Sandwich	RMB 79
Beef Burger	RMB 89

PASTA AND EGGS

Spaghetti Bolognaise	RMB 68
Lasagne Verde Al Forno	RMB 69
Spicy Chicken Pizza	RMB 89
Ham and Pineapple Pizza	RMB 89
Seafood Pizza	RMB 89

MAIN COURSES

Grilled Garlic Sirloin Steak	RMB 168
Vienna Cowboy Steak	RMB 228
Grilled Mutton with Nut and Garlic	RMB 158
Bacon Wrapped Pork Tenderloin	RMB 148

DESSERTS

Croissant	RMB 35
Chiffon Cake	RMB 28
Cookies	RMB 39
Seasonal Fruits with Ice Cream	RMB 39

III Late Night Menu 送餐菜单

送餐菜单
参考译文

Available From 11 a.m. to 11 p.m.

STARTERS AND SALADS

Smoked Norwegian Salmon	RMB 78
Beef Carpaccio	RMB 68
Hearts of Romaine	RMB 59
Seasonal Field Greens	RMB 46

SOUPS

Soup of the Day	RMB 39

SANDWICHES AND BURGERS

Ham and Cheese Sandwich	RMB 78
Beef Burger	RMB 89

PASTA

Spaghetti Bolognaise	RMB 68
Lasagne Verde Al Forno	RMB 69

DESSERTS

Tiramisu	RMB 48
Ice Cream	RMB 35

BEVERAGES

Coffee	RMB 78
Hot Chocolate	RMB 68

Unit 23

Dialogues 对话

I Dialogue 1: In the Kitchen① 在厨房①

A: Good morning, Mr. Hays.

B: Good morning, Mr. Li.

A: Can you tell me what my job is for today?

B: Please select and clean the vegetable. And prepare the ingredients for breakfast.

A: OK. Is there anything else you want me to do?

B: Nothing else, thank you.

New Words

select [sɪ'lekt] vt. 选择；挑选

clean [kliːn] vt. 使……干净

ingredient [ɪn'griːdɪənt] n. （烹调的）原料

breakfast ['brekfəst] n. 早餐

听力在线

参考译文

II Dialogue 2: In the Kitchen② 在厨房②

A: Could you tell me what dishes are for breakfast?

B: They are fried egg, poached egg, toast, bacon, boiled frankfurter, ham, cereal and cheese.

A: And any milk?

B: Of course, more milk should get ready.

A: What are the fried egg with?

B: Cheese, ham, bacon etc.

A: Any other suggestions?

B: No. Thanks for your assistance.

New Words

frankfurter ['fræŋkfɜːtə] n. 熏肠

poached [pəʊtʃt] v. 水煮（荷包蛋）

cereal ['sɪərɪəl] n. 谷类植物；谷物

etc. [ˌɪt'setərə] adv. [拉] 等等

suggestion [sə'dʒestʃ(ə)n] n. 建议，意见

assistance [ə'sɪst(ə)ns] n. 帮助

听力在线

参考译文

III Dialogue 3: In the Kitchen③ 在厨房③

A: What are the chef 's specialities for today?

B: The chef 's specialities are: crepes with crab, smoked trout mousse and tournedos provencal.

A: Have you got ready for the chef 's specialities ingredients?

B: No, how many portions shall I prepare?

A: First please prepare sixty portions.

B: Yes, I see. Please don't worry.

New Words

chef [ʃef] n. 厨师，大师傅

special ['speʃ(ə)l] adj. 特殊的；专门的

speciality [ˌspeʃɪ'ælɪtɪ] n. 特制品

portion ['pɔːʃ(ə)n] n. 一部分

听力在线

参考译文

IV Dialogue 4: In the Kitchen④ 在厨房④

A: Can you tell me what fish is used for stewed cream fish?

B: Perch, sir.

A: The quality of the perch is not high, why not use sole instead?

B: Because we have not sole and only have perch.

A: More cream is needed?

B: Of course. Please don't forget to have some French cheese powder.

A: I see, thank you very much.

B: You are welcome.

New Words

quality ['kwɒlɪtɪ] n. 质量，品质

instead [ɪn'sted] adv. 代替，顶替

听力在线

参考译文

V Dialogue 5: In the Kitchen⑤ 在厨房⑤

A: How much do they pay for each person?

B: Two hundred yuan, excluding drinks.

A: Where are the guests come from?

B: France. Please cook the dishes finely and season them highly, besides every hot dish must stress the odour of wine.

A: What hot dishes are you going to prepare?

B: Here is the menu. Three hot dishes: creamed seafood, fillet steak and roast turkey.

A: What's for the dessert?

B: A plate of fruits and then an ice cream.

A: When shall we begin?

B: Please get ready before 10 o'clock and serve at 11.

New Words

pay [peɪ] v. 付款；支付

excluding [ɪk'sklu:dɪŋ] prep. 除……外，不包括

finely ['faɪnli] adv. 精细地

season ['si:z(ə)n] vt. 调味

besides [bɪ'saɪdz] prep. 除……之外（还有）

stress [stres] n. 强调

odour ['əʊdə] n. 气味

听力在线

参考译文

VI Dialogue 6: In the Kitchen⑥ 在厨房⑥

Cook: Good evening, master. What dishes do we prepare for dinner?

Chef : It's grilled mutton with nut and garlic. It's the dish of the day.

Cook: What method do you use to cook this dish?

Chef : By grilled.

Cook: How long does it take to cook it?

Chef : Cook for about 40 minutes.

Cook: Could you tell me the way to cook it?

Chef : Of course.

New Words

kitchen ['kɪtʃɪn; -tʃ(ə)n] n. 厨房

cook [kʊk] n. 厨师；厨工

master ['mɑ:stə] n. 大师

dishes ['diʃiz] n. 盘（dish 的名词复数）；一盘食物
dinner ['dɪnə] n. 正餐；晚餐
mutton ['mʌt(ə)n] n. 羊肉
nut [nʌt] n. 坚果，坚果果仁
method ['meθəd] n. 方法

听力在线

参考译文

VII Dialogue 7: In the Kitchen⑦ 在厨房⑦

Cook: How about the flavor of this course?
Chef : It tastes too salty.
Cook: Is the sauce enough for it?
Chef : Yes, it's enough. And it tastes much better after adding some sugar.
Cook: I garnish this dish carefully.
Chef : How nice it looks! Now it is all fine in color, odour, flavor and figure.

New Words

flavor ['fleɪvə] n. 味；香料
taste [teɪst] n. 滋味；味觉；风味
salty ['sɔːltɪ; 'sɒ-] adj. 含盐的；咸的
sauce [sɔːs] n. 调味汁；酱汁
add [æd] vt. 增加
sugar ['ʃʊgə] n. 糖；一块[茶匙等] 糖
garnish ['gɑːnɪʃ] vt. 装饰；给……加配菜
carefully ['keəfəlɪ] adv. 仔细地
odour ['əʊdə] n. 气味
figure ['fɪgə] n. 图解；轮廓

听力在线

参考译文

VIII Dialogue 8: Egg Station 鸡蛋烹调台

Cook : Good Morning, Sir!
Guest: Good Morning!
Cook : This is our egg cooking station, We have fried egg, scrambled egg and omelet to choose. Which one would you like?
Guest: Fried egg, Please.
Cook : How would you like your egg done? Sunny side up or over easy?
Guest: Over easy, please.
Cook : Certainly, Sir.

Egg Cooking...Guest waiting...

Cook : It's Ok! (Put the egg into guest's plate.)

Guest: Thank you!

Cook : You are welcome.

New Words

egg [eg] n. 鸡蛋

station ['steɪʃ(ə)n] n. 站，车站；驻地

fried [fraɪd] adj. 油炸的，油煎的

scramble ['skræmb(ə)l] v. 把……搅乱；炒（蛋）

omelet ['ɔmlit] n. 煎蛋饼，煎蛋卷

choose [tʃu:z] v. 选择；进行挑选

sunny side up ['sʌnɪ saɪd ʌp] 煎一面荷包蛋

over easy ['əʊvə 'i:zɪ] v. 两面煎的（一面快煎好时翻到另一面稍煎片刻）

plate [pleɪt] n. 盘子，盆子

听力在线

参考译文

二维码资源

使用说明

《西餐烹饪英语》（第5版）包含纸质教材、听力练习、参考译文及练习答案共三部分学习资源，其中，听力练习、参考译文及练习答案需要扫码学习。

纸质教材涵盖5个模块共23个单元，具体学习时，应当依次通过看图、连线、认菜品说原料、读单词、课后练习，循序渐进、不断强化对西餐烹饪原料和菜品等英语术语的学习。全书为全彩版，西厨房的各种物品、各类西式菜品原料及菜品成品一目了然，您可以形象直观地辨识西餐原材料和西餐用品。

听力练习共包括33个音频学习文件，需要和纸质教材配套使用。前21个练习与教材的前21个单元一一对应，是这些单元中出现的专业术语的跟读练习；后8个练习是Unit23的8个情景对话练习。您通过反复跟读，可在掌握西餐烹调技术的同时，能看懂并熟记各种西餐菜谱、设备用具、调味品和烹饪原材料所对应的英文。听力练习取材专业、语速适中、发音标准，扫码后可随时随地收听练习。每天花5分钟时间，就能轻松搞定西餐烹饪专业英语。

听力练习、在线选择练习（附答案）、参考译文需要先扫码再使用。参考译文主要是Module5中出现的星级酒店西餐菜单译文，以及8个情景对话的中文译文。

参考文献
References

[1] 韦恩·吉斯伦. 专业烹饪[M]. 大连：大连理工大学出版社，2005.

[2] 李廷富. 环球美食：牛肉菜肴精选[M]. 南京：江苏科学技术出版社，2007.

[3] 姜玲. 厨师职业英语[M]. 北京：高等教育出版社，2009.

[4] 蒋湘林. 西式面点制作. 2版[M]. 北京：旅游教育出版社，2022.

[5] 顾健平. 西餐制作. 2版[M]. 北京：旅游教育出版社，2022.

[6] 李秀斌. 现代餐饮英语实务教程[M]. 广州：世界图书出版公司，2005.

[7] 陈培根. 饭店服务英语[M]. 北京：高等教育出版社，1990.

[8] 郭亚东. 西式烹调师[M]. 北京：中国劳动社会保障出版社，2001.